Electropolymerization

Electropolymerization

Editor

Armand Vance

scitus
academics

Electropolymerization
Edited by **Armand Vance**

ISBN: 978-1-68117-293-4
Library of Congress Control Number: 2016937478

www.scitusacademics.com
Box No. 4766, 616 Corporate Way,
Suite 2, Valley Cottage,
NY 10989

Printed in United States of America on Acid Free Paper

Preface

The focus in recent years has been on thin polymer films. These thin polymer films are important for many technological applications, from coatings and adhesives to organic electronic devices, sensors and detectors. Electropolymerization is polymerization in the presence of electric current. Polymerization is any process that chemically combines relatively small molecules called monomers to produce a very large polymer-like chain or network molecule. All the monomer molecules can be the same or two, three or more compounds. Usually, at least 100 monomer molecules must be combined to create a product with unique physical properties, such as elasticity, high tensile strength or the ability to form fibers, which differentiates polymers from smaller and simpler molecules; many thousands of monomer units are often included in a single polymer molecule. Electrochemical polymerization is preferable, especially if the polymer is intended for use as thin polymer films, since electro-generation enables fine film thickness control, an important parameter for the manufacture of equipment. In addition, it has been shown that the material properties can be changed by controlling the electrodeposition process by parameter.

This book covers basic and introductory topics as well as advanced insights. The book provides a timely overview of the recent knowledge of the use of Electropolymerization in the preparation of new materials and the conduction of polymers and varied applications in industry.

Table of Contents

Chapter 1

Biofabrication Using Pyrrole Electropolymerization for the Immobilization of Glucose Oxidase and Lactate Oxidase on Implanted Microfabricated Biotransducers

Christian N. Kotanen [1,2], Olukayode Karunwi [1,2] and Anthony Guiseppi-Elie [1,2,3,4,5]

[1]*Center for Bioelectronics, Biosensors and Biochips (C3B), Clemson University Advanced Materials Center, Anderson, SC 29625, USA*

[2]*Department of Bioengineering, College of Engineering and Science, Clemson University, 301 Rhodes Research Center, Clemson, SC 29634, USA*

[3]*Department of Chemical and Biomolecular Engineering, College of Engineering and Science, Clemson University, 132 Earle Hall, Clemson, SC 29634, USA*

[4]*Holcombe Department of Electrical and Computer Engineering, College of Engineering and Science, Clemson University, 105 Riggs Hall, Clemson, SC 29634, USA*

[5]*ABTECH Scientific, Inc., Biotechnology Research Park, 800 East Leigh Street, Richmond, VA 23219, USA*

ABSTRACT

The dual responsive Electrochemical Cell-on-a-Chip Microdisc Electrode Array (ECC MDEA 5037) is a recently developed electrochemical transducer for use in a wireless, implantable biosensor system for the continuous measurement of interstitial glucose and lactate. Fabrication of the biorecognition membrane via pyrrole electropolymerization and both in vitro and in vivo characterization of the resulting biotransducer is described. The influence of EDC-NHS covalent conjugation of glucose oxidase with 4-(3-pyrrolyl) butyric acid (monomerization) and with 4-sulfobenzoic acid (sulfonization) on biosensor performance was examined. As the extent of enzyme conjugation was increased sensitivity decreased for monomerized enzymes but increased for sulfonized enzymes. Implanted biotransducers were examined in a Sprague-Dawley rat hemorrhage model. Resection after 4 h and subsequent in vitro re-characterization showed a decreased sensitivity from 0.68 (±0.40) to 0.22 (±0.17) $\mu A \cdot cm^{-2} \cdot mM^{-1}$, an increase in the limit of detection from 0.05 (±0.03) to 0.27 (±0.27) mM and a six-fold increase in the response time from 41 (±18) to 244 (±193) s. This evidence reconfirms the importance of biofouling at the bio-abio interface and the need for mitigation strategies to address the foreign body response.

Keywords: Biofabrication; pyrrole; polypyrrole; electropolymerization; biosensors; amperometry; glucose oxidase; glucose; in vivo; hemorrhage

1. INTRODUCTION

The fabrication of multi-analyte biotransducers is a major technical challenge when the length scales of the individual transducer elements are on the order of microns [1,2,3]. Such biotransducers are critical for in vivo (intramuscular) wireless biosensor systems [4] that allow for immediate and continual pre-hospital monitoring of lactate and glucose to inform patient specific interventions that will improve survivability from hemorrhage [2,4]. Continued examination of interstitial compartments using biosensors will aid in understanding the temporal relationships among biomarkers of stress in these environments and how they relate to hemorrhagic shock states [5]. Electropolymerization of conductive polymers such as polypyrrole [6,7,8] and polyaniline [9,10], as well as electropolymerization of non-conductive polymers, such as poly(o-phenylenediamine) [11], poly(m-phenylenediamine) [12,13] and poly(phenylene oxide) [14,15], have been well studied and is one of the emerging additive methods of biofabrication that may be used to guide and deposit biological entities such as enzymes, antibodies, nucleic acids, sub-cellular fragments, and even whole cells to metallic or semi-conducting electrode sites of more complex devices [16,17,18]. Biofabrication techniques using conducting electroactive polymers have been previously described for use in the design and fabrication of the biological recognition membranes of biotransducers [17,19]. Polypyrrole is a conductive electroactive polymer electropolymerized from its monomer pyrrole, which, because of its facility to direct the deposition of dopant and entrained macromolecules onto electrode surfaces, has been previously described and reviewed for the fabrication of amperometric, voltammetric, and impedimetric biotransducers [20,21,22].

Biorecognition membranes of precisely controlled thickness may be fabricated through judicious control of electropolymerization charge density [23]. In so doing, a wide range of copolymers may be synthesized and a wide range of monomeric and polymeric dopants may be incorporated through judicious control of the materials and solvent composition of the electropolymerization bath [24,25]. In addition, post-processing of biorecognition membrane layers, such as over oxidation or chemical derivatization may be used to enhance or stabilize performance of the fabricated bio-layer. Most importantly, all these processing steps may be performed in batch or continuous mode, in bio-benign aqueous environments at pH values close to 7.0 and with the use of modest electrode potentials under conditions that are close to ambient; conditions that are compatible with biological recognition entities and that prevent denaturation. Thus, conducting electroactive polymers (CEPs) have many attractive features such as selectivity of enzyme entrapment to electrode surfaces [26,27], reduction of interfacial impedance, thickness control via charge density [28], enzyme stability, biocompatibility [29,30,31], and adjustment of biosensor enzyme kinetic properties. Polypyrrole, along with other electroactive polymers, once formed, can be surface modified and used for various chemical coupling reactions [32,33]. During electropolymerization, the biological molecule may serve as the dopant anion (most enzymes possess a net negative charge). Hence, biomolecules, such as enzymes, can be used to dope polypyrrole during electropolymerization when fabricating biosensors [17,27,34,35]. Enzymes such as glucose oxidase possess amino acid residues in their primary structure to which covalent coupling of other functional moieties can be performed. For instance GOx from Aspergillus niger (EC 1.1.3.4, PDB ID 1CF3) has 15 lysine residues per subunit (two subunits per molecule of GOx) possessing primary amines available for covalent chemical modification. The covalent chemical modification of an enzyme in such a manner has the potential to influence the extent of its incorporation into the biorecognition membrane layer during the electropolymerization process.

Biotransducers fabricated with electroconductive hydrogels (ECHs) [35,36] offer the tremendous benefits of enzyme stability and protection from proteases in the host environment during in vivo application [35] but with a drawback of suboptimal sensitivity and compromised limits of detection for a wireless biosensor system [2,35]. It is believed that enzyme loading (Units/cm^2) and transport limitations compromise performance. Through enzyme modification that aims to improve incorporation (loading), it should be possible to enhance current densities of polypyrrole-based amperometric biosensors. As a direct means of improving biosensor performance, i.e., increasing biosensor sensitivity and maximum current, lysine groups of enzymes were functionalized with (i) a pyrrole moiety (monomerization) thereby directly incorporating enzymes via co-polymerization into the growing polypyrrole layer, or (ii) a sulfonic acid moiety (sulfonization) thereby incorporating enzymes as better dopants into the growing polypyrrole layer. Both non-covalent and covalent enzyme modifications via monomerization and sulfonization were performed and the influence of these on biosensor performance was studied. Furthermore, the effect of a polypyrrole-polystyrenesulfonate (PPy-PSSA) seeding layer of low charge density (10 mC/cm^2) intended to create a "flash" of highly adherent but otherwise porous PPy on the electrodes was developed as a means to enhance biofabrication and bioanalytical performance of subsequent enzyme layers.

The in vivo performance of a biosensor system [37], based on the MDEA 5037 fabricated with electropolymerized polypyrrole, was examined in a Sprague-Dawley animal model. Previous characterizations of these systems in vitro have been performed and performance parameters of sensitivity, apparent Michaelis-Menten constant, limits of detection, stability, biosensor channel crosstalk, and linear range have been examined [2,27,35,38]. Sensors fabricated with polypyrrole-based electroconductive hydrogels have shown no loss of sensitivity in vitro over an 18-day period when stored in PBS buffer at 4 °C [38]. Glucose biosensors fabricated with overoxidized polypyrrole were observed to limit the response to ascorbic acid to no more than 5% of the total response to glucose, reduce signal response to other negatively charged endogenous interferents by 92% and have an interferent rejection ratio of 12:1 [35,38]. Nonspecific adsorption of enzymes companion sites during the biofabrication process has been shown to be insignificant and to have little to no influence on biosensor response [2]. Dual analyte electrochemical biotransducers for the amperometric detection of glucose and lactate have been developed for use in thin layer flow cells that employ flow injection analysis [39]. The current transducers, developed in conjunction with ABTECH Scientific, Inc. (Richmond, VA, USA), were microlithographically fabricated, decorated with the biorecognition layer, characterized in vitro, and implanted into the trapezius muscle of Sprague Dawley rats. Internal calibration, response under controlled hemorrhage conditions, and post–resection re-characterization were used to evaluate biotransducer performance.

2. EXPERIMENTAL SECTION

2.1. Chemicals and Reagents

Poly(styrene sulfonic acid) (PSSA 30%, MW = 70,000, 6–10% sulfonation, ρ = 1.10 g/mL) was purchased from Polysciences, Inc. Dulbecco's phosphate buffered saline, benzophenone (reagent grade 99%), pyrrole monomer (reagent grade 98+%), 4-(3-pyrrolyl)butyric acid (PyBA), glucose oxidase (GOx, E.C. 1.1.3.4 from Aspergillus niger), β-d(+)-glucose, 1-ethyl-3-[3-dimethylaminopropyl] carbodiimide (EDC), N-hydroxysulfosuccinimide (Sulfo-NHS), N-hydroxysuccinimide (NHS), 4-sulfobenzoic acid (SBA), ferrocene monocarboxylic acid (FcCOOH), and all other common solvents, were purchased from Sigma Aldrich Co. (St. Louis, MO, USA). Pyrrole monomer was purified by double passage through an alumina silicate column (Supelclean™, LC-Alumina-A SPE, 570*2-U, 1 g). Solutions were prepared in deionized water (DI-water) prepared by purifying distilled water through a Milli-Q® plus

(Millipore Inc.) ultrapure water system. The glucose stock solution of 1.0 M was prepared and allowed to mutarotate overnight before use. Phosphate buffered saline (PBS) was used at 0.1 M at pH = 7.2 unless noted otherwise.

2.2. Enzyme Monomerization and Sulfonization

4-(3-pyrrolyl)butyric acid (for monomerization) and 4-sulfobenzoic acid (for sulfonization/dopantization) were covalently conjugated to amine groups of lysine residues of glucose oxidase. For monomerization of enzymes, a solution of 0.175 M Py and 0.025 M PyBA was prepared in DI-water and its pH adjusted to 4.5–5.5 using dropwise addition of 1.0 M NaOH. For sulfonization of enzymes, a solution of 0.2 M Py and 0.625 mM SBA was prepared in DI-water (pH ≈ 4–4.5 without need for adjustment). The acid groups of PyBA or SBA were activated by the addition of 0.625 mM EDC, the concentration needed to achieve a 10:1 mole ratio of PyBA:GOx or SBA:GOx, was incorporated and allowed to incubate for 1 h at 25 °C. Next 0.625 mM of NHS was incorporated and the solution allowed to incubate for 5 min at 25 °C. Finally, 1.0 mg/mL of GOx was incorporated, the solution shaken vigorously and allowed to incubate for 24 h at 4 °C. This process was repeated to achieve a 2:1 mole ratio of PyBA:GOx or SBA:GOx in conjugation and was repeated without EDC-NHS coupling to serve as the non-conjugated controls.

2.3. Electropolymerization and Electrochemical Characterization

Electropolymerization of pyrrole and electrochemical characterization of polypyrrole thin films were performed using a PAR 283 Potentiostat/Galvanostat (Princeton Applied Research) equipped with PowerSuite® software and a BAS-100B/W Electrochemical Analyzer with a BAS PA-1 preamplifier module used to amplify the current and to filter out noise (BASi, West Lafayette, IN, USA). All experiments were carried out in a three-electrode setup with platinum microelectrodes (ϕ = 100 μm; BASi), the Electrochemical Cell-on-a-Chip Microdisc Electrode Arrays (ECC MDEA 5037-Pt; ABTECH Scientific, Inc.) as the working electrode, an Ag/AgCl (3 M KCl) reference electrode (RE803; ABTECH Scientific, Inc.) and a large area platinum mesh counter electrode.

2.4. Transducer Cleaning and Seeding with Polypyrrole

Two types of electrochemical transducers were used; platinum microelectrodes (μE) for the in vitro characterization of polymer films and microlithographically fabricated micro disc electrode arrays (MDEA) for animal implantation. The platinum microelectrode was mechanically polished for 1 min using 1.0 μm diamond slurry and then washed with methanol. This was followed by further polishing for 1 min using 0.05 µm alumina and rinsed with DI-water in order to expose a fresh platinum surface. Electrodes were subsequently ultrasonicated, for 3 min each in DI-water, propanol, DI-water; cleaned in a UV-ozone Cleaner (Boekel Industries, Philadelphia, PA, USA) and cathodically cleaned by repeated cycling over the range 0.0 to −1.2 V in phosphate buffered saline (PBS) vs. Ag/AgCl [40]. The ECC MDEA 5037-Pt is a microlithographically fabricated dual analyte electrochemical transducer intended for amperometric and voltammetric biosensor application. Developed for simultaneous monitoring of interstitial glucose and lactate [1]. The transducer possesses 37 recessed microdiscs arranged in a hexagonal array. Each disc has ϕ = 50 microns for a total working electrode area, WEA = 7.3×10^{-4} cm^2 [41]. The microlithographic fabrication and geometric patterning of the three-electrode electrochemical biotransducers have been previously described [41] and assembly and packaging of the chip in a manner suitable for implantation into small vertebrate animals has similarly been described [42]. In summary, the electrochemical transducers (0.2 cm × 0.4 cm × 0.05 cm) were fabricated from electron beam vapor deposited platinum (100 nm) on an adhesion promoting titanium/tungsten (Ti/W) layer (10 nm) onto a 0.5 mm thick electronics grade borosilicate glass (Schott D263). The electrodes were laid out as two separate three-electrode electrochemical cells and the metallization passivated with 0.5 micron thick silicon nitride (Si_3N_4). The nitride layer was eventually patterned and fluoro-plasma etched to reveal the array of multiple microdiscs of the working electrode (7.3×10^{-4} cm^2) the large area counter electrode (7.3×10^{-3} cm^2) and the shared reference electrode (7.3×10^{-5} cm^2) that were connected to the five bonding pads. The MDEA 5037-Pt transducer was first ultrasonicated for 3 min in each of DI-water, isopropyl alcohol (IPA), and DI-water. Next, the transducer was placed in a UV-ozone cleaner (Boekel Industries) under irradiated ozone generation for 10 min followed by 1 min of ultrasonication in IPA. This was followed by plasma surface modification of the transducer to generate hydroxyl groups on the silicon nitride surface (Harrick Plasma Cleaner). The transducer was then immersed in PBS buffer, made the working electrode of a three electrode electrochemical cell and was cathodically cleaned by sweeping the potential between 0 to −1.2 V vs. Ag/AgCl at 100 mV/s for 40 cycles.

In general, biospecificity was conferred to each array-working electrode of the device by immobilizing a bioreceptor to each working electrode. To support robust biorecognition membrane formation, electrodes were first seeded with a thin electropolymerized film of PSSA doped polypyrrole (PPy-PSSA). This adherent film was formed via potentiostatic electrosynthesis at 750 mV vs. Ag/AgCl to a charge density of 10 mC/cm^2 from an aqueous solution of 0.2 M Py and 0.05 M repeat units of PSSA as dopant prepared in DI-water and pH-adjusted to pH = 4.75 using 0.1 M NaOH [21,35].

2.5. Conferring Biospecificity to Polypyrrole Seeded Transducers

Following covalent or non-covalent conjugation, biospecificity was conferred to the microelectrodes through further electropolymerization in the freshly prepared conjugation solutions to form the covalently conjugated polypyrrole systems of PPy(SBA-con-GOx) and P(Py-co-PyBA-con-GOx). Non-covalently conjugated systems, useful as controls, did not incorporate EDC or NHS in the electropolymerization reaction solutions and these resulted in the following systems of PPy(SBA)-GOx, P(Py-co-PyBA)-GOx. Further control systems (PPy-GOx) were prepared with 0.2 M Py and 1 mg/mL GOx in DI-Water. After preparation, biotransducers were placed in PBS and refrigerated at 4 °C over night to allow for non-specifically adsorbed enzymes to leach out. Biospecificity was then conferred to the working electrode arrays through further electropolymerization in solutions containing 0.175 M Py, 0.025 M PyBA, and 1 mg/mL of GOx or LOx prepared in DI-Water with pH adjustment to 4.5–5.5 using 1 M NaOH. Electropolymerization for entrapping enzymes proceeded at 850 mV vs. Ag/AgCl for 100 mC/cm^2. Before use, enzyme-modified microelectrodes were over-oxidized (OO-PPy) by repeatedly cycling the electrode in PBKCl (100 mM, pH 7.2) between −200 to 1,300 mV vs. Ag/AgCl for 40 cycles.

2.6. In Vitro Biosensor Calibration Prior to in Vivo Implantation

The 8100-K1 fixed frequency wireless dual potentiostat system was used (Pinnacle Technology, Lawrence, KS, USA). The kit contained the Pinnacle 8151 wireless dual potentiostat, the voltage programmer, and a receiver base station (model 8106) with USB cables. Software for data acquisition (PAL) was also included in the kit. Biotransducers were interfaced with the 8151 wireless potentiostat via a custom connector. Two-electrode amperometric measurements of glucose or lactate was enabled, with the on-board working microelectrode array serving as the working electrode and the shorted on-board counter and reference electrodes serving as the counter electrode. A bias potential of 0.65 V was programmed into the 8151 and applied to the working electrode array of the MDEA 5037 with respect to the onboard counter electrode. To ensure functionality prior to implantation, in vitro calibration of MDEA 5037s was performed in PBS buffer at 37 °C, within 8 h, prior to implantation. Steady state amperometric current produced was measured and the sensitivity of the device was determined from the slope of the calibration plot. In vitro calibrated sensitivity (S_o) was determined by the ratio of the resulting change in current density based on electrode geometric area to change in concentration of analyte in buffer in units of $\mu A \cdot cm^{-2} \cdot mM^{-1}$. Time dependent estimations of intramuscular glucose, $G_o(t)$ or lactate, $L_o(t)$ were made based on this reference sensitivity after implantation of the device. The kinetic parameters were determined using Lineweaver-Burk (LWB) analysis of biosensor amperometric dose response. The apparent Michaelis-Menten constant K_{Mapp} and the maximum current response, I_{max}, were calculated from the slope and the intercept of the Lineweaver-Burk plots [20,21]. Limits of detection were calculated by dividing 3*STDev (three times the standard deviation) of the steady-state, blank solution current response by the calculated sensitivity.

2.7. Implantation of Biotransducers into the Trapezius Muscles of Sprague-Dawley Rats

All animal studies were conducted under Clemson University IACUC Protocol # 2011-013, approved on 9 March, 2011, and valid through 8 March, 2014, under the supervision of Clemson University's Institutional Animal Care and Use Committee (IACUC). Sprague-Dawley rats (250–450 g) were housed in plastic cages (1 rat/cage) at the Godley-Snell Research Center at Clemson University and kept on a 12 h light/dark cycle, with food and water provided ad libitum. Sprague-Dawley rats were shaved, sterilized, and placed under anesthesia on a warming pad. Throughout the biosensor implantation and evaluation experiments, the rats were maintained under anesthesia using 3% isoflurane flow. An incision was made above the trapezius muscle on the back to expose the fascia and muscle. Fascia was spread apart using blunt-tipped scissors to expose the muscle. A small pocket was formed by spreading the muscle tissue apart using microtweezers. The MDEA 5037 was carefully inserted into the pocket by hand. The muscle tissue was closed over the working region of the biotransducer and secured with a suture. Fascia was secured over the muscle tissue with a suture and the wound was closed with surgical staples. After implantation, the MDEA 5037 biotransducer was interfaced with the 8151 dual potentiostat and allowed to reach steady state for an hour. The rat was observed during this time to ensure stability.

2.8. In Vivo Biosensor Sensitivity and Response Time to Bolus Infusion of Analytes

After biosensor response reached a steady state baseline measurement of intramuscular analytes, measurements of blood analytes from the saphenous vein were made. Blood glucose was measured using a test-strip based Alpha TRAK glucose meter (Abbott Laboratories). A bolus of solution consisting of 3 M glucose and 0.4 M lactate prepared in 0.1 M PBS (pH = 7.4) was infused, up to 0.5 mL, through the tail vein. Procedures for measurement of interstitial analytes lasted for up to 3 h. One hour was used to allow the biosensors to reach a steady state. Blood analyte measurements and bolus infusions were made after reaching steady state and at 30 min intervals thereafter. Changes in amperometric response deviating from steady state were observed after the bolus injection of glucose and lactate into venous blood. Estimated interstitial analyte responses measured by biosensors calibrated in vitro were compared to analyte concentrations in the blood. Interstitial concentrations of glucose in non-diabetic rats have been shown to have a good 1:1 correlation to plasma glucose concentration at the basal state [43]. Sensitivity was determined based on in vivo responsiveness to exogenously changing levels of glucose and lactate. Rats were subsequently euthanized after procedures and the trapezius muscle surrounding the biotransducer was resected in order to safely recover biotransducers. Resected biotransducers were re-characterized in vitro to assess the functionality and lifetime of the device in vivo. Upon explantation the base current, sensitivity and response time of biosensors were once again evaluated in PBS at 37 °C. Resected biotransducers were considered functional if sensitivity and maximum current were maintained, being not significantly different from freshly prepared sensors, can detect the physiologically relevant range of glucose or lactate, and have a response time of approximately 1 min.

3. RESULTS AND DISCUSSION

3.1. Effects of a Polypyrrole Seeding Layer on Analytical Performance of Biotransducers

To create effective, stable and reproducible biotransducers, it is necessary to establish an intimate and stable interface between the biorecognition membrane layer and the underlying physicochemical transducer, in this case, the metallic electrode. To achieve this, an approach has been developed that uses a nano-dimensioned seeding layer of electropolymerized, pristine, polypyrrole-polystyrene sulfonate. Electropolymerization serves as an additive fabrication process that places an organic electronic layer in intimate contact with an electrified metallic layer. No other surfaces of the device are affected. The effects of seeding transducers with PPy-PSSA prior to subsequent electropolymerization and entrapment of enzymes was explored. A low charge density of 10 mC/cm^2, corresponding to a film of 25 nm [28], was used as a seeding layer prior to applying 100 mC/cm^2 of PPy/GOx. Unseeded biotransducers, serving as a control, had a sensitivity of 0.77 (±0.35) $\mu A \cdot cm^{-2} \cdot mM^{-1}$, an I_{max} of 16.8 (±6.5) $\mu A \cdot cm^{-2}$, a linear dynamic range maximum of 15.8 (±2.8) mM, a K_{Mapp} of 11.1 (±4.3) mM, and a limit of detection of 0.07 (±0.05) mM. Seeded biotransducers had a sensitivity of 1.44 (±0.78) $\mu A \cdot cm^{-2} \cdot mM^{-1}$, an I_{max} of 35.3 (±18.3) $\mu A \cdot cm^{-2}$, a linear dynamic range maximum of 19.4 (±1.0) mM, a K_{Mapp} of 11.8 (±0.3) mM, and a limit of detection of 0.04 (±0.03) mM.

Figure 1 summarizes the effects of the PPy-PSSA seeding layer at 10 mC/cm^2 on the enzyme kinetic parameters of biotransducers fabricated at 100 mC/cm^2.

Both the sensitivity and the apparent Michaelis-Menten enzyme kinetic parameters were statistically equivalent suggesting no deleterious influence of the adhesion promoting seeding layer on biotechnical performance. Both of the biotransducer types had much greater sensitivity to glucose in comparison to hydrogel coated biotransducers which have been reported to have a sensitivity of 0.045 $\mu A \cdot cm^{-2} \cdot mM^{-1}$ [35]. Previous characterizations of unseeded immobilization of GOx with polypyrrole have yielded biotransducers with glucose sensitivity of approximately 0.6 $\mu A \cdot cm^{-2} \cdot mM^{-1}$ which is similar to the results obtained in this body of work [38]. The linear range of an amperometric biosensor system is typically a function of analyte diffusivity through the bioactive membrane [44]. The linear dynamic range of seeded biotransducers was not significantly broader than that of unseeded biotrasducers. The nano-thin PPy-PSSA layer was not enough to serve as a barrier to diffusion of enzymatically generated hydrogen-peroxide in order to increase the linear range or adversely affect other enzyme kinetic parameters. The K_{Mapp} values determined show that no changes to enzyme affinity occurred as a result of the seeding layer. Seeded biotransducers showed less variability in K_{Mapp} indicating good reproducibility of enzyme affinity compared to unseeded transducers. For amperometric, enzyme-based biosensors, the maximum current is related to total enzyme activity [17]. The nearly 2-fold increase in maximum current was statistically significant ($p = 0.045$). These results are a good indication that the seeding layer serves to enhance loaded enzyme activity. It is likely that the seeding layer provides a surface that facilitates enzyme adsorption/entrapment without denaturation. Polypyrrole is known to have a stabilizing influence on enzymes [45]. Enzymes being immobilized onto an unseeded electrode will likely be entrapped within a mechanically disparate interface of Pt|PPy that may not preserve the structure of immobilized enzymes as effectively.

The principal role of the seeding layer was to significantly decrease the average biofabrication time from 68 (±22) min for the unseeded biotransducer to 35 (±6) min for the seeded biotransducer ($p = 0.007$); a reduction of 50%, and to improve reproducibility in biofabrication. Electropolymerization efficiency is an important biofabrication parameter. Enzymes require a hydrated, protective environment in order to function in vivo. Electropolymerization of pyrrole to entrap GOx within polypyrrole to yield a biosmart, pHEMA-based hydrogel has been used to fabricate in vivo biotransducers [46]. These hydrogels provided the necessary protective hosting of the immobilized biomolecules [47] and the in vivo biocompatibility for the protection of the biotransducers [35]. However, hydrogel modified transducers require considerable overpotential for electropolymerization; as much as 1,000 mV vs. Ag/AgCl and the polymerization times are comparable to that of unmodified electrodes [35].

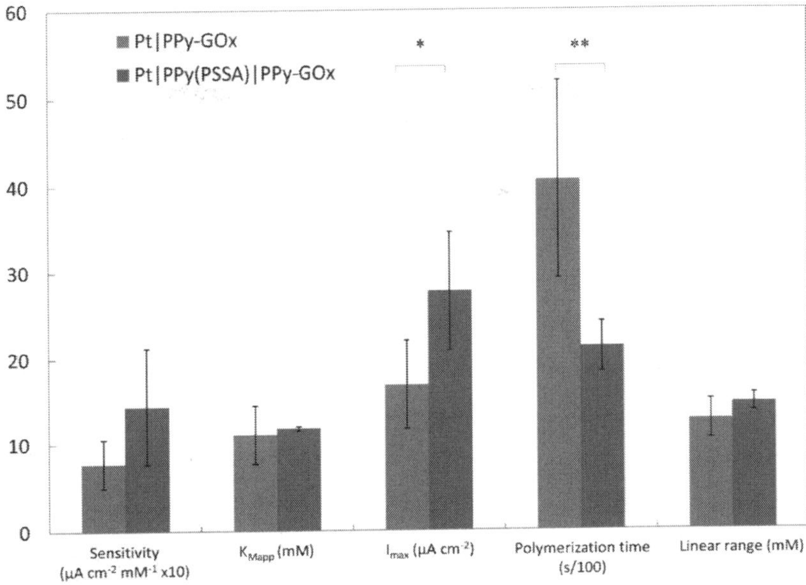

Figure 1. The effects of the PPy-PSSA seeding layer of 10 mC/cm^2 on the enzyme kinetic parameters of biotransducers fabricated using 100 mC/cm^2 (error bars show 95% confidence intervals, $n \geq 5$). * = $p < 0.05$; ** = $p < 0.01$.

3.2. Enzyme Monomerization and Sulfonization

The monomerization of enzymes [48,49,50,51] and biomolecules [52,53] is a well-practiced art and the sulfonization of GOx using PEG spacer arms to a poly(2-acrylamido-2-methylpropane sulfonic acid) has likewise recently been described [54,55]. In this work, the amine groups of lysine residues of GOx were covalently coupled to either 4-(3-pyrrolyl)butyric acid or 4-sulfobenzoic acid using EDC-NHS coupling chemistry, Figure 2. The amperometric dose response of biotransducers to glucose was measured for (i) covalently conjugated systems: **PPy(SBA-con-GOx)**, **P(Py-co-PyBA-con-GOx)**, (ii) non-covalently conjugated systems: **PPy(SBA)-GOx**, **P(Py-co-PyBA)-GOx**, and (iii) control systems: **PPy-GOx**. The covalently conjugated systems with a 10:1 molecular ratio of PyBA:GOx or SBA:GOx are depicted in Figure 3.

The approach to covalent coupling considered in this work is intended to be a simple and effective means of enzyme conjugation that can be performed in a single solution and immediately used for electropolymerization without need for sophisticated separation steps. Moreover, these reactions are non-targeted, being potentially reactive with amine groups near the active site and with the potential for compromise of global enzyme structure and activity. It is noted that not all lysine groups will be available for conjugation and that the total number of lysine groups conjugated may likely be less than that of the feed ratio. Monomerized enzymes were prepared with the intent of having them be directly incorporated into the backbone of the forming electropolymerized polypyrrole. Sulfonized enzymes were prepared in order to have them act as more effective dopants and so be more efficiently electrostatically entrapped within the growing layer of positively charged polypyrrole backbones.

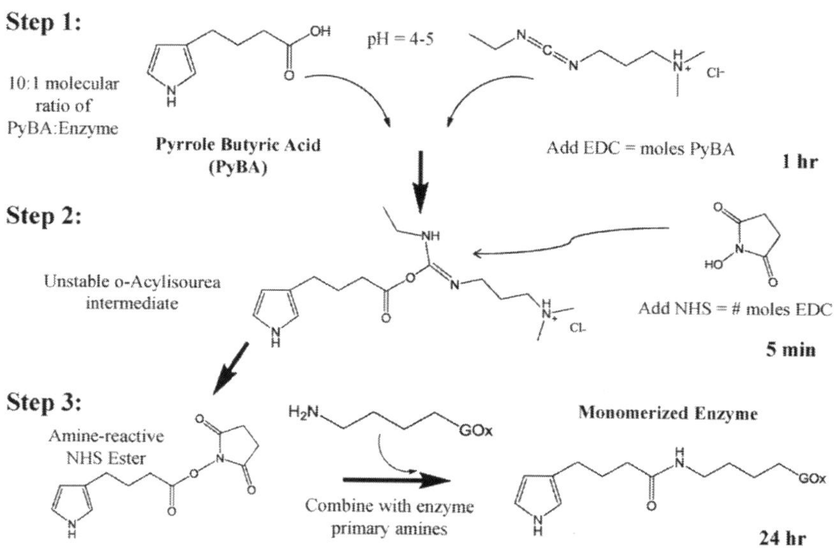

Figure 2. Scheme of enzyme modification using EDC-NHS coupling chemistry to monomerize glucose oxidase with PyBA; Sulfonization of enzymes with SBA was performed in a similar fashion.

3.3. Electropolymerization Kinetics of Covalently Conjugated Systems

The kinetics of electropolymerization, defined as the time to reach 100 mC/cm^2 of charge, were compared for all systems studied. The control systems of PPy-GOx had an electropolymerization time of 68 ± 22 min. Enzyme systems containing PyBA, namely P(Py-co-PyBA)-GOx and P(Py-co-PyBA-con-GOx), both had significantly greater ($p < 0.03$) electropolymerization times compared to PPy-GOx. Enzyme systems PPy(SBA)-GOx, PPy(SBA-con-GOx) had significantly lower electropolymerization times compared to all other systems ($p < 0.02$). Sulfobenzoic acid is a highly mobile dopant capable of screening charge and making electropolymerization kinetics more facile for polypyrrole formation [56]. No significant difference in kinetics was noted between covalent and non-covalent conjugation for either PyBA or SBA systems. A decrease in variability in electropolymerization time was however observed for the covalently conjugated systems when compared to the non-covalently conjugated systems. There is likely some competition for doping sites within the formed polypyrrole between adsorbed glucose oxidase and PyBA/SBA in the non-covalent systems, whereas for covalently conjugated systems, the dopant becomes solely the adsorbed monomerized or sulfonated enzyme.

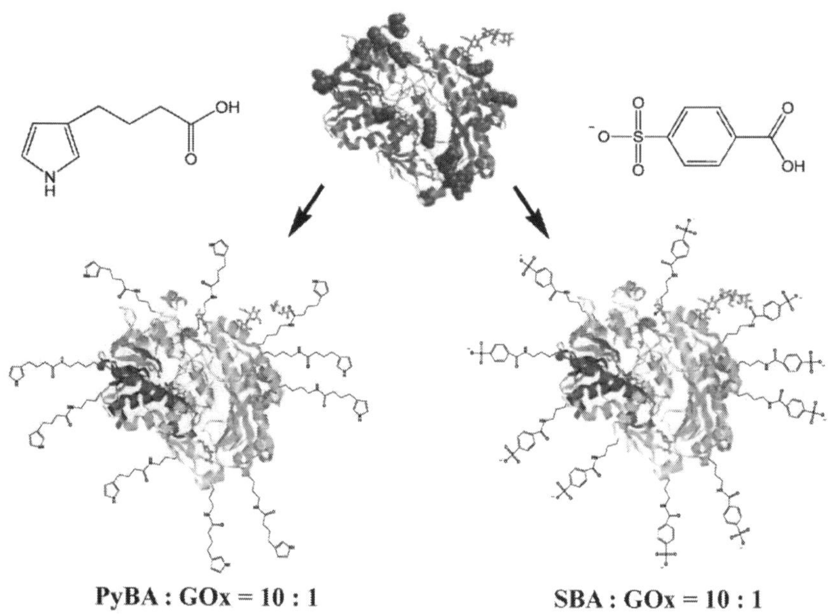

Figure 3. Subunit of glucose oxidase from Aspergillus niger with Lysine groups in red (**top**), monomerized enzymes used for preparing P(Py-co-PyBA-con-Gox) (**bottom left**) and sulfonated (**bottom right**) enzymes used for preparing PPy(SBA-con-GOx) with 10:1 ratio of PyBA or SBA conjugation to enzymes.

3.4. Characterization of Enzyme Conjugation using Ferrocene Monocarboxylic Acid

Cyclic voltammetry in the presence of a redox mediator, ferrocene monocarboxylic acid, was used to probe and confirm the extent of monomerization and sulfonization of enzymes by EDC-NHS coupling. A redox mediator, ferrocene monocarboxylic acid, was conjugated to glucose oxidase to a saturating extent using a 30:1 conjugation ratio of FcCOOH:GOx to confirm the effectiveness of covalent conjugation. In order to separate unreacted components of the solution after conjugation had taken place, samples were allowed to dialyze in a Slide-A-Lyzer dialysis cassette (Thermo Scientific Pierce, Rockford IL) in deionized water over night in a water bath, which was changed with fresh deionized water every 3 h. These samples are denoted as 'Dialyzed'. After dialysis was complete the solutions were immediately characterized electrochemically. Cyclic voltammograms obtained at platinum microelectrodes in the presence of DI-water, unmodified GOx enzyme and covalently conjugated FcCONH-GOx enzyme is shown in Figure 4. The formal potential for ferrocene monocarboxylic acid is approximately 320 mV vs. Ag/AgCl [57]. Higher anodic and cathodic currents were observed for FcCOOH-GOx conjugates compared to just enzyme or DI-water. The oxidation and reduction peaks decreased for the conjugates after dialysis due to the removal of unreacted FcCOOH. Redox activity was still evident, however, indicating that covalent conjugation had occurred. Response from 1 mg/mL of GOx showed no changes and was identical to samples of GOx and FcCOOH mixed together without EDC/NHS and dialyzed (data not shown).

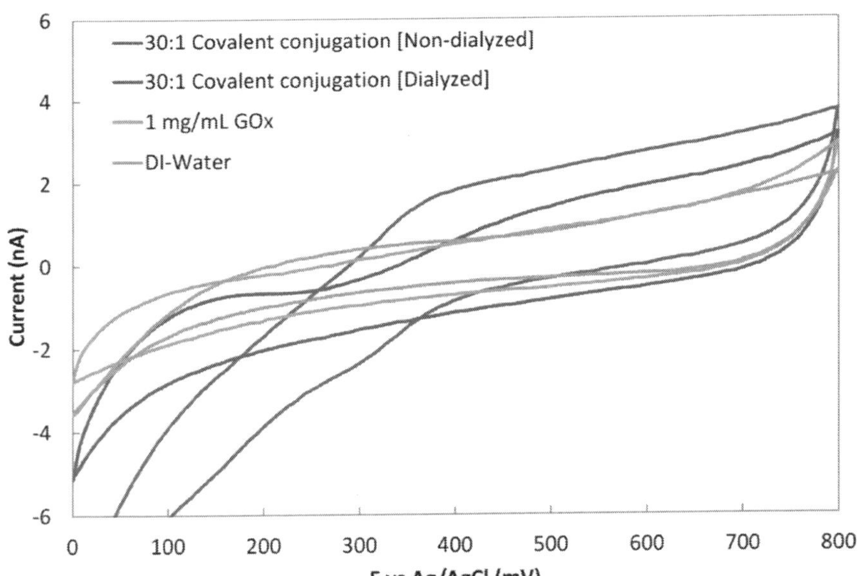

Figure 4. Cyclic voltammograms of GOx and FcCOOH-GOx covalent conjugates in DI-water at a scan rate of 50 mV/s.

3.5. Bioanalytical Performance of Monomerized and Sulfonated Enzyme Biosensors

The dose response curves of biotransducers fabricated using monomerized GOx and sulfonated GOx are shown in Figure 5 and biosensor performance parameters for all systems are summarized in Table 1. The covalently sulfonated system, PPy(SBA-con-GOx), had a sensitivity of 1.92 ± 0.2 μA·cm^{-2}·mM^{-1}, a response time of 18 ± 5 s, a detection limit of 2.2 × 10^{-3}± 1.8 × 10^{-3} mM, a K_{Mapp} of 11.1 ± 5.6 mM, a maximum current of 39.2 ± 11.7 μA·cm^{-2} and a linear dynamic range of 10.2 ± 3.2 mM. The PPy(SBA-con-GOx) system was significantly more sensitive ($p < 0.0002$) than the PPy-GOx system. The PPy(SBA-con-GOx) system was shown to have the best limits of detection with significantly better ($p = 0.002$) resolution compared to P(Py-co-PyBA-con-GOx). Response times were equivalent among all systems. The non-covalently coupled SBA system, PPy(SBA)-GOx, had a sensitivity of 1.8 ± 1.2 μA·cm^{-2}·mM^{-1} which was found to be the same as PPy-GOx controls. Electropolymerized polypyrrole glucose biotransducers fabricated with GOx conjugated to poly(2-acrylamido-2-methylpropane sulfonic acid) polyanion dopants have been shown to have sensitivity ranging from 180–270 nA/mM/cm^2 with a linear range of 20–26 mM and a Michaelis-Menten constant of 46.7 mM [55,58]. Overall the covalent sulfonization process successfully yielded biotransducers with higher sensitivity but at the cost of decreased linear dynamic range when compared to the GOx-polyanion system found in the literature. Biotransducers were able to maintain unchanged functionality for up to 8 days with storage in PBS at 4 °C. After the 8-day period background current increased and sensitivity decreased during characterization in PBS (data not shown).

Table 1. Dose response (PBS, pH = 7.2, 25 °C) performance parameters for biotransducers fabricated with (10:1 ratio PyBA/SBA:GOx) covalently modified enzymes, non-covalently modified enzymes and control systems having no modification (±standard deviation is shown).

[E] = 1.0 mg/mL	PPy-GOx (n = 4)	PPy(SBA)-GOx (n = 4)	PPy(SBA-*con*-GOx) (n = 2)	P(Py-*co*-PyBA)-GOx (n = 2)	P(Py-*co*-PyBA-*con*-Gox) (n = 2)
Polymerizaton time (R.L. 100 mC/cm²) (min)	68 ± 22	23 ± 19	15 ± 5	204 ± 37	230 ± 132
Sensitivity (μA·cm^{-2}·mM^{-1})	0.87 ± 0.04	1.8 ± 1.2	1.9 ± 0.2	4.8 ± 0.9	0.99 ± 0.28
Response time (s)	41 ± 21	17 ± 9	18 ± 5	17 ± 5	14 ± 5
Detection limit (mM)	0.5 ± 0.3	0.01 ± 0.01	2.2 × 10^{-3} ± 1.8 × 10^{-3}	0.05 ± 0.03	0.02 ± 0.01
K_{Mapp} (mM)	14.1 ± 3.1	8.2 ± 2.6	11.1 ± 5.6	7.2 ± 2.6	19.7 ± 3.9
I_{max} (μA·cm^{-2})	23.3 ± 3.1	30.6 ± 18.8	39.2 ± 11.7	77 ± 25	28.9 ± 4.0
Linear range (mM)	17.4 ± 7.0	9.4 ± 2.6	10.2 ± 3.2	9.5 ± 2.8	25.6 ± 5.6

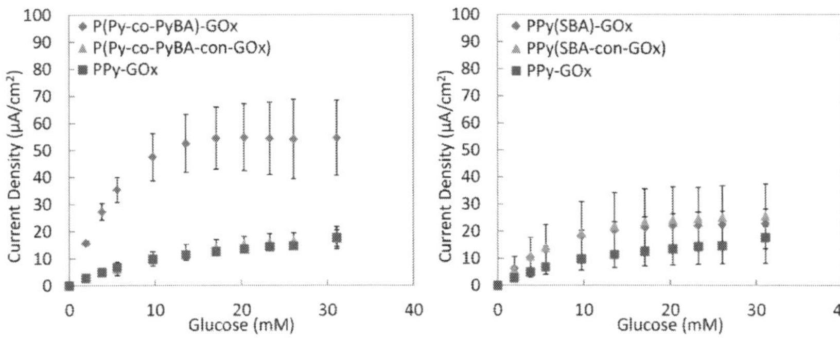

Figure 5. Dose response curves of (**left**) monomerized and (**right**) sulfonated enzyme biotransducers prepared on Pt|PPy-PSSA seeding layers of 10 mC/cm^2 at ϕ = 100 μ·Pt·μE.

The non-covalently conjugated system, P(Py-co-PyBA)-GOx, had a sensitivity of 4.8 ± 0.9 μA·cm^{-2}·mM^{-1}, a response time of 17 ± 5 s, a detection limit of 0.05 ± 0.03 mM, a K_{Mapp} of 7.2 ± 2.6, a maximum current of 77 ± 25 μA·cm^{-2} and a linear range of 9.5 ± 2.8 mM. Its sensitivity was significantly greater than every other biotransducer ($p < 0.045$) except the sulfonated system, PPy(SBA-con-GOx). P(Py-co-PyBA)-GOx had the highest maximum current of all systems studied and was significantly greater ($p < 0.01$) than the PPy-GOx system. Covalently monomerized GOx did not perform statistically different from PPy-GOx. Glucose oxidase that is lysyl modified with an N-substituted 1H-Pyrrole-1-propionic acid has been shown to have higher electrode activity compared to native GOx in polypyrrole membranes [59]. Chemical modification of GOx lysine groups with 3-carboxymethyl pyrrole has also been shown to exhibit higher enzyme activity and storage stability than systems using N-substituted pyrrole [60]. Our results show that incorporation of PyBA without covalent coupling is sufficient for increasing biosensor performance with respect to higher sensitivity and overall enzyme activity without the need to modify electrode geometry or properties. Enhancing surface area of the electrode through use of nanofeatures can improve sensitivity by an order of magnitude, 10 μA·cm^{-2}·mM^{-1} [61]. This improvement, however, may not carry over to the final in vivo biosensor system which will be coated with bioactive hydrogels that is expected to limit diffusion [35,62].

The MDEA 5037 system was equal in performance with respect to response time and improved in performance with respect to sensitivity when compared to wire-type epoxy-polyurethane implantable systems which have been shown to have response times ranging from 20–200 s with sensitivity ranging from 1–6 nA/mM [63]. Rapid response is necessary for these implantable systems in order to most accurately report physiological conditions in vivo. For enzyme-based amperometric biosensors, the response time will generally be limited by the diffusion of glucose and hydrogen peroxide through the bioactive polymer membrane. Spin casting hydrogels to control thickness of bioactive membranes can be used to ensure that effective response times (<60 s) are maintained. Linearity of response for these enzyme-modified systems was generally similar to that of biosensors fabricated with GOx entrapped by composites of nanoparticles and polypyrrole with limits between 4 and 13 mM [64,65,66]. Linear range limits of 20 mM have been achieved using PPy coated gold nanoparticles, an increase which comes with decreased sensitivity by a factor of 6 and increased K_{Mapp} [67]. Similar types of observations were made for the covalently monomerized biotransducer system, wherein sensitivity decreased by an order of about 4–5 while increases in K_{Mapp} and linear range occurred. A high linear range is valuable for accurate measurement of high concentrations of glucose without the need for estimation from curve fits. Carbon nanotube doped polypyrrole, when used to entrap GOx onto a glassy carbon electrode, can reach a linear range of up to 50 mM but with compromised sensitivity of 2.3 nA/mM [68].

3.6. Effect of Extent of Monomerizaton or Sulfonation

The extent of monomerization and sulfonation and its impact on biosensor sensitivity was examined by varying the ratio of PyBA and SBA conjugated to glucose oxidase, Figure 6. A ratio of 2:1 was compared to 10:1. Sensitivity decreased as conjugation increased for monomerized enzymes and increased for sulfonated enzymes. A significant decrease in sensitivity from 4.8 ± 0.9 $\mu A \cdot cm^{-2} \cdot mM^{-1}$ at 0:1 conjugation to 0.1 ± 0.01 $\mu A \cdot cm^{-2} \cdot mM^{-1}$ at 2:1 conjugation ($p = 0.02$) and 1.0 ± 0.3 $\mu A \cdot cm^{-2} \cdot mM^{-1}$ at 10:1 conjugation ($p = 0.03$) was noted for the monomerized systems. For sulfonated systems sensitivity significantly increased from 1.0 ± 0.03 $\mu A \cdot cm^{-2} \cdot mM^{-1}$ at 0:1 conjugation to 1.7 ± 0.5 $\mu A \cdot cm^{-2} \cdot mM^{-1}$ at 2:1 conjugation ($p = 0.03$) and 1.9 ± 0.2 $\mu A \cdot cm^{-2} \cdot mM^{-1}$ at 10:1 conjugation ($p < 0.0002$). Monomerized GOx systems had sluggish electropolymerization kinetics with long times required to reach 100 mC/cm^2 of charge density. This would allow a greater number of enzymes to adsorb onto the forming polypyrrole during the polymerization process resulting in a higher sensitivity for the system.

The conjugation to PyBA is likely affecting the active sites of the enzymes leading to the overall decreased sensitivity. The sensitivity increased after increasing the conjugation from 2:1 to 10:1, which may be due to more enzymes being directly incorporated into the polypyrrole network through the enzyme conjugation process. Sulfonated enzymes had the most rapid electropolymerization kinetics, the result of the SO_3^- highly effective dopant. The monotonically increasing sensitivity for sulfonated enzymes suggests that electrostatic interactions are likely playing a key role in enzyme immobilization via electropolymerization. The contents of the enzyme-monomer solution used for electropolymerization will largely dictate the polymer structure, morphology, thickness, conductance, and biosensor performance parameters [69,70]. The non-covalent and covalent conjugated enzymes in this work were designed with the intent to maximize biotransducer enzyme content by having the enzymes dope the forming polypyrrole. This was shown to have a significant influence on the kinetics of polymerization and improvements in sensitivity, maximum current, and linear range. However, these improvements were achieved with limited reproducibility. The same charge density was maintained throughout PPy film production but was likely limited and affected by the low current densities observed during fabrication, which would have caused differences in film thickness and morphology. Optimization of the electrolyte or dopant composition of the electropolymerization solution will be necessary in the future to improve the quality of the polypyrrole films produced and thereby improve the bioanalytical reproducibility among biotransducers. For the preliminary characterization of intramuscularly implanted biosensor systems each biotransducer was characterized and calibrated individually prior to use.

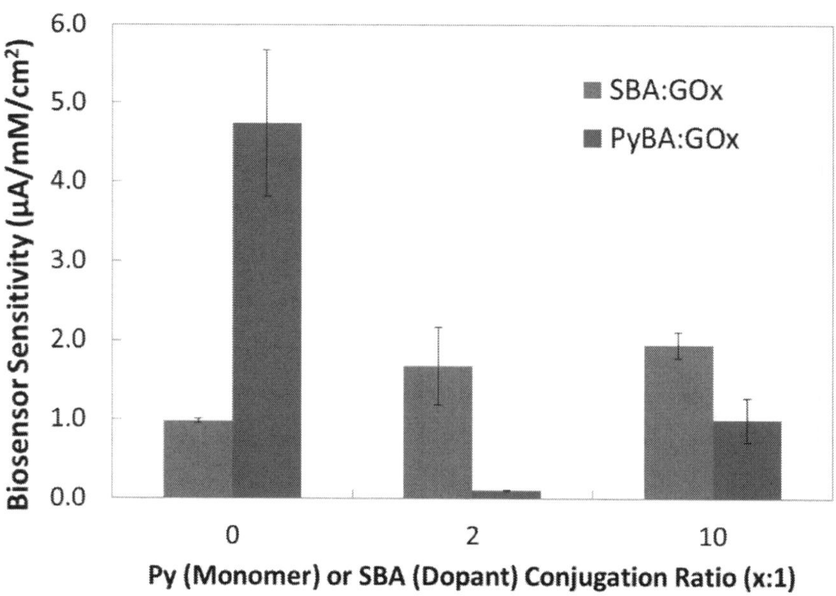

Figure 6. Effects of the extent of monomerization and sulfonization via increasing molecular ratio of PyBA or SBA to GOx on biosensor sensitivity.

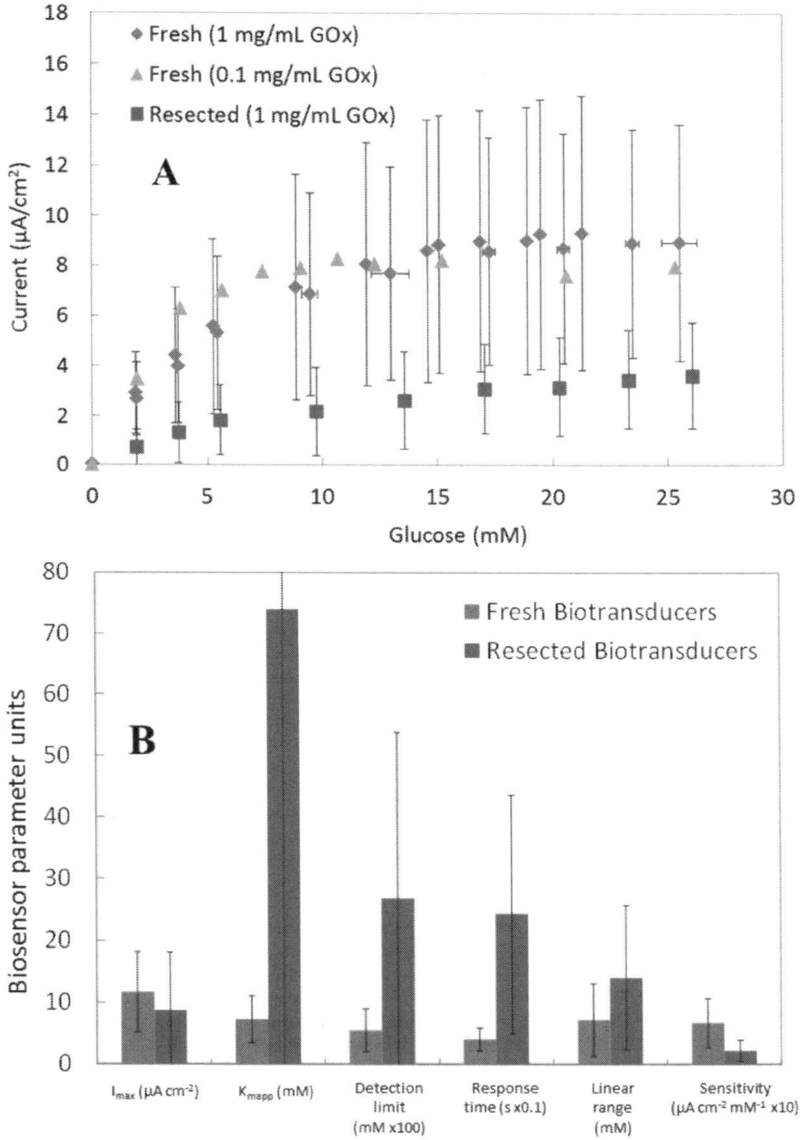

Figure 7. (**A**) Dose response of MDEA 5037-Pt biotransducers to glucose with enzyme concentration used for fabrication shown in parentheses, and (**B**) Bioanalytical performance of glucose biosensors after being freshly fabricated and following resection from rat trapezius muscle (650 mV bias, PBS buffer, pH = 7.2, 25 °C, ± standard deviation is shown).

3.7. Performance of Freshly Prepared and Resected Biotransducers

Biotransducers fabricated using the MDEA 5037-Pt platform were characterized immediately after their fabrication and again immediately following resection from the trapezius muscle of Sprague-Dawley rats. Figure 7 shows the dose–response curves and performance parameters of glucose biosensors fabricated using glucose oxidase, an FAD-based enzyme, before and after resection. The biosensors varied in terms of their performance and, thus, were calibrated independently prior to use in vivo. Table 2 summarizes the key bioanalytical performance characteristics of the biotransducers before and after resection. The MDEA 5037 biotransducers were implanted in muscle tissue for up to 3 h. The overall performance of resected biotransducers was diminished compared to when they were freshly prepared which was expected.

Table 2. Bioanalytical performance parameters for lactate and glucose responsive biotransducers, freshly fabricated and following resection after 3 h of implantation in the trapezius muscle of Sprague-Dawley rats (650 mV bias, PBS, pH = 7.2, 25 °C, ±standard deviation is shown).

	MDEA 5037-Pt|PPy(PSSA)|p(Py-*co*-PyBA)-LOx [E] = 1.0 mg/mL		MDEA 5037-Pt|PPy(PSSA)|p(Py-*co*-PyBA)-GOx [E] = 1.0 mg/mL	
	Freshly prepared biotransducers (n = 3)	Resected biotransducers (n = 3)	Freshly prepared biotransducers (n = 4)	Resected biotransducers (n = 4)
Sensitivity ($\mu A \cdot cm^{-2} \cdot mM^{-1}$)	0.36 ± 0.15	-	0.68 ± 0.40	0.22 ± 0.17
Response time (s)	86 ± 26	-	41 ± 18	244 ± 193
Detection limit (mM)	7.9 ± 5.6	-	0.05 ± 0.03	0.27 ± 0.27
K_{Mapp} (mM)	−9.1 ± 15.8	-	7.3 ± 3.4	74 ± 130
I_{max} ($\mu A \cdot cm^{-2}$)	−1.9 ± 5.8	-	11.7 ± 6.5	8.8 ± 9.3
Linear range (mM)	6.6 ± 0.9	-	7.2 ± 5.9	14.1 ± 11.6

Lactate biotransducers did not produce adequate current densities in response to in vitro lactate challenge to be considered for in vivo analysis when paired with the Pinnacle 8151 wireless dual-potentiostat. The lactate sensors had a good sensitivity of 0.36 ± 0.15 $\mu A \cdot cm^{-2} \cdot mM^{-1}$ and a linear dynamic range of 6.6 ± 0.9 mM. Based on the observed I_{max} and K_{Mapp} the lactate biosensors did not adhere well to L-B analysis. Given the detection limits of 7.9 ± 5.6 mM, currently fabricated lactate biotransducers could only accurately discern changes in interstitial lactate corresponding to extremes of hyperlactatemia (Δ[lactate] > 4 mM) [71,72]. Glucose biotransducers showed diminished sensitivity after use in vivo, a decrease from 0.68 ± 0.40 to 0.22 ± 0.17 $\mu A \cdot cm^{-2} \cdot mM^{-1}$ ($p = 0.08$), and an increase in response time, which increased from 41 ± 18 to 244 ± 193 s ($p = 0.08$). The detection limit increased from 0.05 ± 0.03 to 0.27 ± 0.27 mM ($p = 0.17$), K_{Mapp} increased from 7.3 ± 3.4 to 74 ± 130 mM ($p = 0.35$), I_{max} decreased from 11.7 ± 6.5 to 8.8 ± 9.3 ($p = 0.30$), and the linear range increased from 7.2 ± 5.9 to 14.1 ± 11.6 ($p = 0.34$).

Implanted biotransducers may demonstrate compromised performance due to component-based and biocompatibility-based failures [73]. The following mechanisms of failure for the explanted biosensors were analyzed: lead detachments, electrical shorts, component delamination and degradation, biofouling, electrode passivation, and fibrous encapsulation. Biotransducers were exposed to whole blood and the possibility of protein adsorption, as well as the reactive oxygen species of the acute inflammatory response during their period of implantation. The dramatic increase in response time and K_{Mapp} is likely due to the effect of proteins being adsorbed from surrounding tissue, plasma, and interstitial fluid to the biotransducer surface. Protein adsorption has been shown to have a negative influence on glucose diffusion for implantable and continuous biosensor systems [74,75]. However, no physical deterioration of the MDEA 5037 transducer was observed and none of the resected biotransducers had spurious responses indicating that neither instrumentation problems nor transducer compromises were causing the changes in performance. Limits of detection were maintained at adequate levels to discern changes in glucose using the biosensor system. Production of hydrogen peroxide has been identified as the main agent for short-term enzyme degradation in vivo [76]. The lack of significant change in maximum current shows that the original enzyme activity was still present in the system after short-term use in vivo, supporting these previous findings. Overall, the MDEA 5037 biotransducers maintained their activity after short-term implantation with respect to sensitivity, maximum current, range of response, and response time.

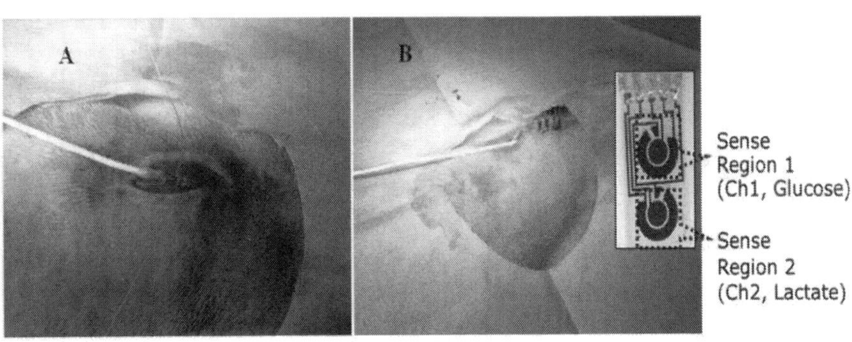

Figure 8. Sprague-Dawley rat under anaesthesia with, (**A**) the MDEA 5037 biotransducer implanted into trapezius muscle and secured into place with a suture, and (**B**) the incision closed and secured around the MDEA 5037 with surgical staples.

3.8. In Vivo Response of MDEA 5037 Biotransducers to Exogenously Changing Analytes

Three dual responsive MDEA 5037-Pt|PPy(PSSA)|p(Py-co-PyBA)-GOx(Ch1)-Pt|PPy(PSSA)|p(Py-co-PyBA)-LOx(Ch2) biotransducers were tested in vivo (MDEA-1, MDEA-2 and MDEA-3). Figure 8 shows the implanted biotransducer within the trapezius muscle of Sprague-Dawley rats under anesthesia.

Following implantation, devices were allowed to stabilize and steady state currents were established and sustained for approximately 1 h. The average blood glucose and amperometric in vivo responses of the glucose sensitive channels are shown in Figure 9. MDEA-1 and MDEA-3 were prepared having a glucose sensitive cell (Channel 1) and a lactate sensitive cell (Channel 2). Response of Channel 2 for these biotransducers was omitted for clarity. MDEA-2 utilized two glucose sensitive cells (Channels 1 and 2) to compare differences in response to glucose on the same biotransducer. Intramuscular glucose via the biosensor was compared to venous blood glucose collected from the saphenous vein and measured with the Alpha TRAK. The first sensor, MDEA-1, came to a stable base-line current, and interstitial glucose concentrations did not vary much between 1.91–2.23 mM. This seemed reasonable as blood glucose did not vary, remaining between 5.49–5.88 mM. One of the sensors (MDEA-2, Channel 2) showed a sustained increase in interstitial glucose after tail vein infusions of glucose. For the case of MDEA-2, both channels were sensitive to glucose. However, only one of the channels (Channel 2) showed a change in interstitial glucose while the other (Channel 1) did not. MDEA-3 showed a decreasing trend of interstitial glucose concentration with a rapid increase and subsequent fall within minutes after a tail vein infusion.

Glucose measurements made using the intramuscularly implanted MDEA 5037 biosensors and the whole blood Alpha TRAK handheld were compared, Figure 10. Intramuscular glucose measurements were consistently lower than blood glucose measurements. Previous comparisons of blood versus interstitial glucose concentrations in humans have indicated that as blood glucose increases interstitial values will increase at a slower rate with a total lag of 17.2 ± 7.2 min [77]. Glucose concentrations have been shown to change significantly faster in the blood of anesthetized rats with a rate of 6.8 ± 2.0 mg/dL/min compared to subcutaneous fluid having a rate of 3.9 ± 1.3 mg/dL/min [78]. For rats given an infusion of insulin the difference between subcutaneous glucose and blood glucose has been shown to vary from 30% to 120% within the first 30 min [78]. Thus, we can expect to see major differences between compartments in the anesthetized rats receiving large boluses of glucose and lactate. We should expect to see some similarities between blood and interstitial glucose concentrations at the basal level. Tissue contact after implantation likely played a role in response after implantation. Short-duration implantations, such as these, do not provide sufficient time for tissue remodelling to allow sustained

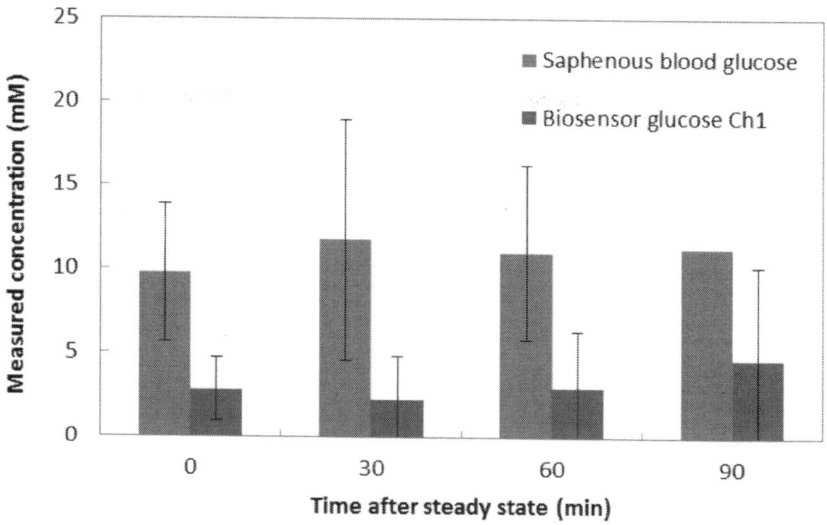

Figure 9. Average whole blood and intramuscular (biosensor, 650 mV) glucose response at 30 min intervals after reaching steady state in the trapezius muscle of MDEA-1 in a male Sprague-Dawley rat of 412 g, MDEA-2 in a male Sprague-Dawley rat of 408 g and MDEA-3 in a female Sprague-Dawley rat of 265 g (infusions were of 3 M glucose in PBS, error bars show standard deviation of n ≥ 2).

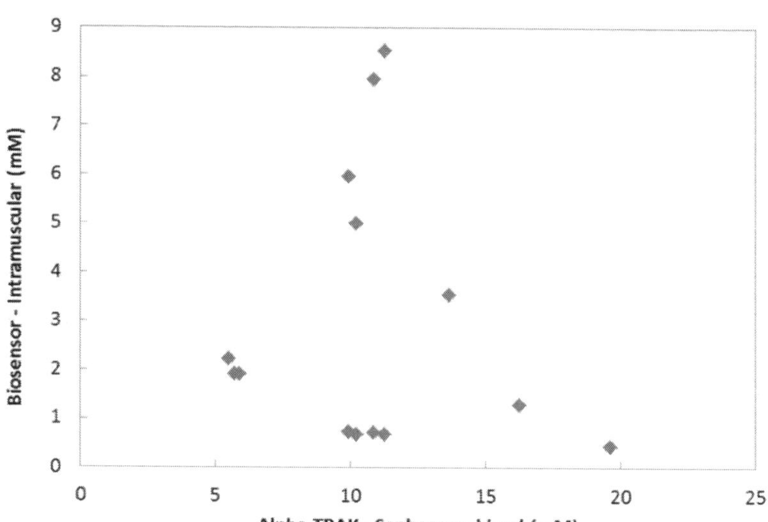

Figure 10. Response comparison of the implanted MDEA 5037 glucose biotransducers with respect to hand-held Alpha TRAK measurement of venous blood glucose as a reference.

intimate contact between the host tissues and the biotransducer surface [79,80]. Longer implantation times will be incorporated in the future. In addition to this, a more gradual ramp administration of glucose will be used. This technique has been shown to yield more rapid and accurate sensor response and a more physiologically relevant response in animals [80,81]. Another drawback of this preliminary study is that the diet, size, and gender of the rats were not controlled.

Background current could not be fully accounted for with these biotransducers as both electrochemical channels were used for simultaneous dual analyte detection. Accuracy in continuous monitoring can be improved by ratiometrically correcting for background current [82]. Background current stability of biosensors is known to display drift in vivo [83]. Future studies will need to take into consideration background current immediately prior to implantation. Electroconductive hydrogel modified biotransducers will also be used in future studies, which will likely have better stability in vivo compared to just polypyrrole modified biotransducers. The biosmart hydrogels have been shown to mitigate protein adsorption [62] which may lead to better short-term performance in vivo. The inclusion of hydrogels will be necessary for the protection of enzymes and mitigation of fibrous encapsulation for long-term implantation of biosensors.

4. CONCLUSIONS

Platinum microelectrodes seeded with a PPy-PSSA layer prior to additive electropolymerization of a PPy-enzyme layer had no significant effect on biotransducer sensitivity, linear dynamic range, limits of detection and response time but served to increase maximum current and reduce biofabrication time. These results indicate that entrapping enzymes onto polypyrrole seeded electrodes better maintains activity compared to entrapment directly onto the electrodes. Seeded transducers also had significantly more facile polypyrrole electropolymerization kinetics. The seeding layer may be a good option for making the polymerization kinetics more facile; with a smaller overpotential in these hydrogel systems when highly mobile dopants (Cl^-, HPO_4^{2-}) are not available. Sulfonization of enzymes was shown to be a fast and an effective means of improving biotransducer performance. Sulfonization using sulfobenzoic acid at a 10:1 SBA:GOx ratio was shown to be useful in making electropolymerization kinetics more facile and doubling biotransducer sensitivity compared to native enzyme systems. Monomerization of enzymes using pyrrole butyric acid at a 10:1 PyBA:GOx ratio did not significantly influence biotransducer performance, however, non-covalently incorporated pyrrole butyric acid dramatically increased biotransducer sensitivity and decreased linear dynamic range. This increase may be due to an increased number of adsorbed enzymes during electropolymerization.

The MDEA 5037 biotransducers and Pinnacle 8151 wireless dual potentiostat biosensor system was successfully prepared and characterized in vivo. Both, glucose and lactate biosensors were prepared and characterized in vitro prior to implantation. Lactate biosensor performance was insufficient for characterization in vivo as the limit of detection was not low enough. Glucose biosensors performed as expected based on previous characterization in vitro. MDEA 5037 biotransducers maintained full functionality after 3 h of implantation in the trapezius muscle of Sprague-Dawley rats. No significant decreases in sensitivity and maximum current were observed after biotransducers were resected. Protein adsorption onto the MDEA 5037 biotransducer surface after short-term use in vivo is likely the cause for dramatically increased response time and increased K_{Mapp}. To improve both short-term and long-term performance in vivo, the biotransducers will need to be coated with a biocompatible hydrogel layer that can mitigate protein adsorption. Likewise, background current will also need to be reduced.

ACKNOWLEDGMENTS

This work was supported by the US Department of Defense (DoDPRMRP) grant PR023081/DAMD17-03-1-0172, by the Consortium of the Clemson University Center for Bioelectronics, Biosensors and Biochips (C3B) and by ABTECH Scientific, Inc.

REFERENCES

1. Park, J.; Kim, C.S.; Choi, M. Oxidase-coupled amperometric glucose and lactate sensors with integrated electrochemical actuation system. IEEE Trans. Instrum. Meas. **2006**, 55, 1348–1355.

2. Kotanen, C.; Guiseppi-Elie, A. Characterization of a wireless potentiostat for integration with a novel implantable biotransducer. IEEE Sens. J. **2013**, 14, 768–776.

3. Zimmermann, S.; Fienbork, D.; Flounders, A.W.; Liepmann, D. In-device enzyme immobilization: Wafer-level fabrication of an integrated glucose sensor. Sens. Actuator B Chem. **2004**, 99, 163–173.

4. Endo, H.; Yonemori, Y.; Hibi, K.; Ren, H.; Hayashi, T.; Tsugawa, W.; Sode, K. Wireless enzyme sensor system for real-time monitoring of blood glucose levels in fish. Biosens. Bioelectron. **2009**, 24, 1417–1423.

5. Uyehara, C.F.T.; Sarkar, J. Role of vasopressin in maintenance of potassium homeostasis in severe hemorrhage. Am. J. Physiol. Regul. Integr. Comp. Physiol. **2013**, 305, R101–R103.

6. Bartlett, P.N.; Whitaker, R.G. Electrochemical immobilisation of enzymes: Part I. Theory. J. Electroanal. Chem. Interfacial Electrochem. **1987**, 224, 27–35.

7. Bartlett, P.N.; Whitaker, R.G. Electrochemical immobilisation of enzymes: Part II. Glucose oxidase immobilised in poly-N-methylpyrrole. J. Electroanal. Chem. Interfacial Electrochem. **1987**, 224, 37–48.

8. Sadki, S.; Schottland, P.; Brodie, N.; Sabouraud, G. The mechanisms of pyrrole electropolymerization. Chem. Soc. Rev. **2000**, 29, 283–293.

9. Yehezkeli, O.; Yan, Y.-M.; Baravik, I.; Tel-Vered, R.; Willner, I. Integrated oligoaniline-cross-linked composites of Au nanoparticles/glucose oxidase electrodes: A generic paradigm for electrically contacted enzyme systems. Chem. Eur. J. **2009**, 15, 2674–2679.

10. Bai, J.; Beyer, S.; Trau, D. 3.331. Conjugated polymers for biosensor devices. In Comprehensive Biomaterials; Ducheyne, P., Ed.; Elsevier: Oxford, UK, 2011; pp. 529–556.

11. Palmisano, F.; Centonze, D.; Zambonin, P.G. An in situ electrosynthesized amperometric biosensor based on lactate oxidase immobilized in a poly-o-phenylenediamine film: Determination of lactate in serum by flow injection analysis. Biosens. Bioelectron. **1994**, 9, 471–479.

12. Zhou, H.; Chen, H.; Luo, S.; Chen, J.; Wei, W.; Kuang, Y. Preparation and bioelectrochemical responses of the poly(m-phenylenediamine) glucose oxidase electrode. Sens. Actuator B Chem. **2004**, 101, 224–230.

13. Badea, M.; Curulli, A.; Palleschi, G. Oxidase enzyme immobilisation through electropolymerised films to assemble biosensors for batch and flow injection analysis. Biosens. Bioelectron. **2003**, 18, 689–698.

14. Bartlett, P.N.; Caruana, D.J. Electrochemical immobilization of enzymes. Part V. Microelectrodes for the detection of glucose based on glucose oxidase immobilized in a poly(phenol) film. Analyst **1992**, 117, 1287–1292.

15. Bartlett, P.N.; Caruana, D.J. Electrochemical immobilization of enzymes. Part VI. Microelectrodes for the detection of L-lactate based

on flavocytochrome b2 immobilized in a poly(phenol) film. Analyst **1994**, 119, 175–180.

16. Tseng, T.T.-C.; Yao, J.; Chan, W.-C. Selective enzyme immobilization on arrayed microelectrodes for the application of sensing neurotransmitters. Biochem. Eng. J. **2013**, 78, 146–153.

17. Karunwi, O.; Wilson, A.N.; Kotanen, C.N.; Guiseppi-Elie, A. Engineering the abio-bio interface to enable more than moore in functional bioelectronics. J. Electrochem. Soc. **2013**, 160, B60–B65.

18. Apetrei, C.; Rodríguez-Méndez, M.L.; De Saja, J.A. Amperometric tyrosinase based biosensor using an electropolymerized phosphate-doped polypyrrole film as an immobilization support. Application for detection of phenolic compounds. Electrochimica Acta **2011**, 56, 8919–8925.

19. Palmisano, F.; Zambonin, P.G.; Centonze, D. Amperometric biosensors based on electrosynthesised polymeric films. Fresenius J. Anal. Chem. **2000**, 366, 586–601.

20. Guiseppi-Elie, A. Chemical and Biological Sensor Devices Having Electroactive Polymer Thin Films Attached to Mirofabricated Devices and Possessing Immobilized Indicator Moieties. U.S. Patent No. 5,766,934, 1998.

21. Guiseppi-Elie, A.; Brahim, S.; Wilson, A.M. Biosensors based on electrically conducting polymers. In Handbook of Conducting Polymers: Conjugated Polymer Processing and Applications, 3rd ed.; Skotheim, T., Reynolds, J.R., Eds.; Taylor and Francis: New York, NY, USA, 2007; Volume 2, pp. 12:11–12:45.

22. Cosnier, S. Biomolecule immobilization on electrode surfaces by entrapment or attachment to electrochemically polymerized films. A review. Biosens. Bioelectron. **1999**, 14, 443–456.

23. Valaski, R.; Ayoub, S.; Micaroni, L.; Hümmelgen, I. Influence of film thickness on charge transport of electrodeposited polypyrrole thin films. Thin Solid Films **2002**, 415, 206–210.

24. Lee, H.; Yang, H.; Kwak, J. Dependence of the electrochemical behavior of poly(N-phenylpyrrole) films on the type of anion and solvent used in the electropolymerization. J. Phys. Chem. B **1999**, 103, 6030–6035.

25. Lee, H.; Yang, H.; Kwak, J. Effects of dopant anions and N-substituents on the electrochemical behavior of polypyrrole films in propylene carbonate solution. Electrochem. Comm. **2002**, 4, 128–133.

26. Tseng, T.T.-C.; Monbouquette, H.G. Implantable microprobe with arrayed microsensors for combined amperometric monitoring of the

neurotransmitters, glutamate and dopamine. J. Electroanal. Chem. **2012**, 682, 141–146.

27. Kotanen, C.; Guiseppi-Elie, A. Development of an implantable biosensor system of physiological status monitoring during long duration space exploration. Gravit. Space Biol. **2011**, 23, 55–64.

28. Holdcroft, S.; Funt, B.L. Preparation and electrocatalytic properties of conducting films of polypyrrole containing platinum microparticulates. J. Electroanal. Chem. Interfacial Electrochem. **1988**, 240, 89–103.

29. George, P.M.; Lyckman, A.W.; LaVan, D.A.; Hegde, A.; Leung, Y.; Avasare, R.; Testa, C.; Alexander, P.M.; Langer, R.; Sur, M. Fabrication and biocompatibility of polypyrrole implants suitable for neural prosthetics. Biomaterials **2005**, 26, 3511–3519.

30. Wang, X.; Gu, X.; Yuan, C.; Chen, S.; Zhang, P.; Zhang, T.; Yao, J.; Chen, F.; Chen, G. Evaluation of biocompatibility of polypyrrole in vitro and in vivo. J. Biomed. Mater. Res. A **2004**, 68A, 411–422.

31. Ramanaviciene, A.; Kausaite, A.; Tautkus, S.; Ramanavicius, A. Biocompatibility of polypyrrole particles: An in vivo study in mice. J. Pharm. Pharmacol. **2007**, 59, 311–315.

32. Kang, E.; Neoh, K.; Zhang, X.; Tan, K.; Liaw, D. Surface modification of electroactive polymer films by ozone treatment. Surf. Interface Anal. **1996**, 24, 51–58.

33. Goddard, J.M.; Hotchkiss, J. Polymer surface modification for the attachment of bioactive compounds. Progr. Polymer Sci. **2007**, 32, 698–725.

34. Kotanen, C.N.; Guiseppi-Elie, A. Bioactive electroconductive hydrogels yield novel biotransducers for glucose. Macromol. Symp. **2012**, 317–318, 187–197.

35. Kotanen, C.N.; Tlili, C.; Guiseppi-Elie, A. Bioactive electroconductive hydrogels: The effects of electropolymerization charge density on the storage stability of an enzyme-based biosensor. Appl. Biochem. Biotechnol. **2012**, 166, 878–888.

36. Guiseppi-Elie, A. Electroconductive hydrogels: Synthesis, characterization and biomedical applications. Biomaterials **2010**, 31, 2701–2716.

37. Kotanen, C.N.; Moussy, F.G.; Carrara, S.; Guiseppi-Elie, A. Implantable enzyme amperometric biosensors. Biosens. Bioelectron. **2012**, 35, 14–26.

38. Kotanen, C.N.; Tlili, C.; Guiseppi-Elie, A. Amperometric glucose biosensor based on electroconductive hydrogels. Talanta **2013**, 103, 228–235.

39. Palmisano, F.; Rizzi, R.; Centonze, D.; Zambonin, P.G. Simultaneous monitoring of glucose and lactate by an interference and cross-talk free dual electrode amperometric biosensor based on electropolymerized thin films. Biosens. Bioelectron. **2000**, 15, 531–539.

40. Guiseppi-Elie, A.; Wilson, A.M.; Tour, J.M.; Brockmann, T.W.; Zhang, P.; Allara, D.L. Specific immobilization of electropolymerized polypyrrole thin films onto interdigitated microsensor electrode arrays. Langmuir **1995**, 11, 1768–1776.

41. Guiseppi-Elie, A.; Brahim, S.; Slaughter, G.; Ward, K. Design of a subcutaneous implantable biochip for monitoring of glucose and lactate. IEEE Sens. J. **2005**, 5, 345–355.

42. Rahman, A.R.A.; Justin, G.; Guiseppi-Elie, A. Towards an implantable biochip for glucose and lactate monitoring using micro-disc electrode arrays (MDEAs). Biomed. Microdevices **2009**, 11, 75–85.

43. Aussedat, B.; Dupire-Angel, M.; Gifford, R.; Klein, J.C.; Wilson, G.S.; Reach, G. Interstitial glucose concentration and glycemia: Implications for continuous subcutaneous glucose monitoring. Am. J. Physiol. Endocrinol. Metabol. **2000**, 278, E716–E728.

44. Emr, S.A.; Yacynych, A.M. Use of polymer films in amperometric biosensors. Electroanalysis **2005**, 7, 913–923.

45. Fortier, G.; Belanger, D. Characterization of the biochemical behavior of glucose oxidase entrapped in a polypyrrole film. Biotechnol. Bioeng. **2004**, 37, 854–858.

46. Guiseppi-Elie, A. An implantable biochip to influence patient outcomes following trauma-induced hemorrhage. Anal. Bioanal. Chem. **2011**, 399, 403–419.

47. Guiseppi-Elie, A.; Koch, L.; Finley, S.H.; Wnek, G.E. The effect of temperature on the impedimetric response of bioreceptor hosting hydrogels. Biosens. Bioelectron. **2011**, 26, 2275–2280.

48. Shoemaker, S.G.; Hoffman, A.S.; Priest, J.H. Synthesis and properties of vinyl monomer/enzyme conjugates. Conjugation of l-asparaginase with N-succinimidyl acrylate. Appl. Biochem. Biotechnol. **1987**, 15, 11–24.

49. Wolowacz, S.E.; Yon Hin, B.F.Y.; Lowe, C.R. Covalent electropolymerization of glucose oxidase in polypyrrole. Anal. Chem. **1992**, 64, 1541–1545.

50. Ahuja, T.; Mir, I.A.; Kumar, D.; Rajesh. Biomolecular immobilization on conducting polymers for biosensing applications. Biomaterials **2007**, 28, 791–805.

51. Yon-Hin, B.F.Y.; Smolander, M.; Crompton, T.; Lowe, C.R. Covalent electropolymerization of glucose oxidase in polypyrrole. Evaluation of methods of pyrrole attachment to glucose oxidase on the performance of electropolymerized glucose sensors. Anal. Chem. **1993**, 65, 2067–2071.

52. Lee, J.Y.; Schmidt, C.E. Pyrrole-hyaluronic acid conjugates for decreasing cell binding to metals and conducting polymers. Acta Biomaterialia **2010**, 6, 4396–4404.

53. DeVolder, R.J.; Zill, A.T.; Jeong, J.H.; Kong, H. Microfabrication of proangiogenic cell-Laden alginate-g-Pyrrole hydrogels. Biomaterials **2012**, 33, 7718–7726.

54. Sung, W.J.; Bae, Y.H. A glucose oxidase electrode based on polypyrrole with polyanion/PEG/enzyme conjugate dopant. Biosens. Bioelectron. **2003**, 18, 1231–1239.

55. Sung, W.J.; Bae, Y.H. Glucose oxidase, lactate oxidase, and galactose oxidase enzyme electrode based on polypyrrole with polyanion/PEG/enzyme conjugate dopant. Sens. Actuator B Chem. **2006**, 114, 164–169.

56. Akhtar, P.; Too, C.; Wallace, G.G. Detection of haloacetic acids at conductive electroactive polymer-modified microelectrodes. Anal. Chim. Acta **1997**, 341, 141–153.

57. Cass, A.E.; Davis, G.; Francis, G.D.; Hill, H.A.O.; Aston, W.J.; Higgins, I.J.; Plotkin, E.V.; Scott, L.D.; Turner, A.P. Ferrocene-mediated enzyme electrode for amperometric determination of glucose. Anal. Chem. **1984**, 56, 667–671.

58. Sung, W.J.; Bae, Y.H. A glucose oxidase electrode based on electropolymerized conducting polymer with polyanion-enzyme conjugated dopant. Anal. Chem. **2000**, 72, 2177–2181.

59. Yon-Hin, B.F.; Smolander, M.; Crompton, T.; Lowe, C.R. Covalent electropolymerization of glucose oxidase in polypyrrole. Evaluation of methods of pyrrole attachment to glucose oxidase on the performance of electropolymerized glucose sensors. Anal. Chem. **1993**, 65, 2067–2071.

60. Yon-Hin, B.; Lowe, C. An investigation of 3-functionalized pyrrole-modified glucose oxidase for the covalent electropolymerization of enzyme films. J. Electroanal. Chem. **1994**, 374, 167–172.

61. Xu, G.; Lv, J.; Zheng, Z.; Wu, Y. Polypyrrole (PPy) Nanowire Arrays Entrapped with Glucose Oxidase Biosensor for Glucose Detection. In Proceedings of the 7th IEEE International Conference on of Nano/Micro Engineered and Molecular Systems (NEMS), Kyoto, Japan, 5–8 March 2012; pp. 511–514.

62. Abraham, S.; Brahim, S.; Ishihara, K.; Guiseppi-Elie, A. Molecularly engineered p(HEMA)-based hydrogels for implant biochip biocompatibility. Biomaterials **2005**, 26, 4767–4778.

63. Yu, B.; Long, N.; Moussy, Y.; Moussy, F. A long-term flexible minimally-invasive implantable glucose biosensor based on an epoxy-enhanced polyurethane membrane. Biosens. Bioelectron. **2006**, 21, 2275–2282.

64. Tsai, Y.-C.; Li, S.-C.; Liao, S.-W. Electrodeposition of polypyrrole-multiwalled carbon nanotube-glucose oxidase nanobiocomposite film for the detection of glucose. Biosens. Bioelectron. **2006**, 22, 495–500.

65. Li, J.; Lin, X. Glucose biosensor based on immobilization of glucose oxidase in poly(o-aminophenol) film on polypyrrole-Pt nanocomposite modified glassy carbon electrode. Biosens. Bioelectron. **2007**, 22, 2898–2905.

66. Kong, T.; Chen, Y.; Ye, Y.; Zhang, K.; Wang, Z.; Wang, X. An amperometric glucose biosensor based on the immobilization of glucose oxidase on the ZnO nanotubes. Sens. Actuator B Chem. **2009**, 138, 344–350.

67. German, N.; Ramanavicius, A.; Voronovic, J.; Ramanaviciene, A. Glucose biosensor based on glucose oxidase and gold nanoparticles of different sizes covered by polypyrrole layer. Colloid Surface Physicochem. Eng. Aspect **2012**, 413, 224–230.

68. Wang, J.; Musameh, M. Carbon-nanotubes doped polypyrrole glucose biosensor. Anal. Chim. Acta **2005**, 539, 209–213. [Google Scholar] [CrossRef]

69. Demoustier-Champagne, S.; Ferain, E.; Jérôme, C.; Jérôme, R.; Legras, R. Electrochemically synthesized polypyrrole nanotubules: effects of different experimental conditions. Eur. Polymer J. **1998**, 34, 1767–1774.

70. Chen, X.; Issi, J.P.; Devaux, J.; Billaud, D. Chemically oxidized polypyrrole: Influence of the experimental conditions on its electrical conductivity and morphology. Polymer Eng. Sci. **1995**, 35, 642–647.

71. Levy, B. Lactic acidosis and hyperlactatemia. In Yearbook of Intensive Care and Emergency Medicine; Vincent, J.-L., Ed.; Springer: Berlin/Heidelberg, Germany, 2006; Volume 2006, pp. 88–98.

72. Callaway, D.W.; Shapiro, N.I.; Donnino, M.W.; Baker, C.; Rosen, C.L. Serum lactate and base deficit as predictors of mortality in normotensive elderly blunt trauma patients. J. Trauma **2009**, 66, 1040–1044.

73. Wisniewski, N.; Reichert, M. Methods for reducing biosensor membrane biofouling. Colloid Surface B Biointerface **2000**, 18, 197–219.

74. Wilson, G.S.; Gifford, R. Biosensors for real-time in vivo measurements. Biosens. Bioelectron. **2005**, 20, 2388–2403.

75. Rigby, G.; Ahmed, S.; Horseman, G.; Vadgama, P. In vivo glucose monitoring with open microflow-influences of fluid composition and preliminary evaluation in man. Anal. Chim. Acta **1999**, 385, 23–32.

76. Valdes, T.; Moussy, F. In vitro and in vivo degradation of glucose oxidase enzyme used for an implantable glucose biosensor. Diabetes Tech. Therapeut. **2000**, 2, 367–376.

77. Kulcu, E.; Tamada, J.A.; Reach, G.; Potts, R.O.; Lesho, M.J. Physiological differences between interstitial glucose and blood glucose measured in human subjects. Diabetes Care **2003**, 26, 2405–2409.

78. Schmidtke, D.W.; Freeland, A.C.; Heller, A.; Bonnecaze, R.T. Measurement and modeling of the transient difference between blood and subcutaneous glucose concentrations in the rat after injection of insulin. PNAS **1998**, 95, 294–299.

79. Koschwanez, H.E.; Reichert, W.M. In vitro, in vivo and post explantation testing of glucose-detecting biosensors: Current methods and recommendations. Biomaterials **2007**, 28, 3687–3703.

80. Updike, S.J.; Shults, M.C.; Rhodes, R.K.; Gilligan, B.J.; Luebow, J.O.; von Heimburg, D. Enzymatic glucose sensors: Improved long-term performance in vitro and in vivo. ASAIO J. **1994**, 40, 157–163.

81. Armour, J.C.; Lucisano, J.Y.; McKean, B.D.; Gough, D.A. Application of chronic intravascular blood glucose sensor in dogs. Diabetes **1990**, 39, 1519–1526.

82. Youssef, J.E.; Castle, J.R.; Engle, J.M.; Massoud, R.G.; Ward, W.K. Continuous glucose monitoring in subjects with type 1 diabetes: Improvement in accuracy by correcting for background current. Diabetes Tech. Therapeut. **2010**, 12, 921–928.

83. Jeong, R.-A.; Hwang, J.Y.; Joo, S.; Chung, T.D.; Park, S.; Kang, S.K.; Lee, W.-Y.; Kim, H.C. In vivo calibration of the subcutaneous amperometric glucose sensors using a non-enzyme electrode. Biosens. Bioelectron. **2003**, 19, 313–319.

Chapter 2

Nanocomposites Based on Conducting Polymers and Functionalized Carbon Nanotubes with Different Dopants Obtained by Electropolymerization

Florina Branzoi[1*], Viorel Branzoi[2]

[1]*Institute of Physical Chemistry "Ilie Murgulescu" 202 Splaiul Independenţei, Bucharest, Romania*

[2]*Department of Applied Physical Chemistry and Electrochemistry, University Politehnica of Bucharest, Bucharest, Romania*

ABSTRACT

Nanocomposite films based on poly (3,4-ethylenedioxy thiophene) (PEDOT), functionalized singlewalled carbon nanotubes and different dopants were studied. It was fabricated by a simple oxidative electropolymerization method. The dopant substances used were SDS (Sodium dodecyl sulfate) and tiron (1,2-Dihydroxybenzene-3,5-disulfonic acid disodium salt hydrate). These nanocomposite films were grown electrochemically from aqueous solutions such that constituents were deposited simultaneously onto substrate electrode. The synthetic, morphological and electrical properties of the nanocomposite films obtained were compared. Scanning electron microscopy (SEM) revealed that the composite films consisted of nanoporous networks of SWCNTS (single-walled carbon nanotubes) coated with polymeric film. Cyclic voltammetry (CV), electro-chemical impedance spectroscopy (EIS) and FT-IR spectroscopy demonstrated that these composite films had similar electrochemical response rates to pure polymeric films but a lower resistance and much improved mechanical integrity. The negatively charged functionalized carbon nanotubes (CNTSF) served as anionic dopant during the electropolymerization to synthesize polymer/CNTSF composite films. The specific electrochemical capacitance of the composite films is a significantly higher value than that for pure polymer films prepared similarly. Using these composite films, the modified electrodes with improved properties were obtained. (In this paper, for simplicity, the SWANTs-COOH group will be noted CNTsF which means functionalized carbon nanotubes.)

Keywords: Nanocomposite Films, Functionalized Carbon Nanotubes, SEM, Capacitance, Real Impedance

1. INTRODUCTION

As a result of the first report concerning the preparation of carbon nanotubes (CNTs)/polymer composite like that many efforts have been made to combine CNTs and polymers to produce functional composite materials with superior properties [1] -[4] . Apart from possible improvements in the mechanical and electrical properties of polymers, the formation of CNTs/polymer composites have been and still are explored as a promising approach for an effective incorporation of CNTs into practical devices [5] -[9] . Composite material based on the coupling of conducting organic polymers (CPs) and CNTs have shown that they possess properties of the individual components with a synergistic effect [9] . In this context, a special attention has been paid to the following CPs: polyaniline (PANI), polypyrrole (PPY), polythiophene (PTh) [10] -[13] etc. Among the numerous materials devised, development of the polythiophene derivative, polyethylenedioxythiophene (PEDOT), has shown significant promise for the challenge at hand [14] .

The combination of CNTs with CPs offers an attractive route to reinforce the polymer as well as to introduce electronic properties based on morphological modification or electronic interaction between the two components [4] -[9] [15] -[18] . However, it is difficult to process CNTs and insoluble in most solvents. In order to broaden their applications, it is necessary to tailor their solubility properties. For this reason, in this study, SWCNTs (singlewalled carbon nanotubes) covalently functionalized with polycarboxylic acids, were used. The SWCNTs graft copolymer has excellent solubilities in water and some organic solvents and it also exhibits an order of magnitude increase in electrical conductivity. It was also demonstrated that SWCNTs-COOH (single walled carbon nanotubes functionalized with polycarboxylic acids) showed an improved sensor performance compared to unfunctionalized SWCNTs because polycarboxylic acid is an conductive organic compound in its own right. Additionally, the presence of numerous functional groups in SWCNTs-COOH means that there is potential for covalent immobilization of various big molecules and specially biomolecules.

The chemical functionalization of CNTs (carbon nanotubes) has been shown to increase their solubility in different solvents and facilitate their processing. It has also been demonstrated that the chemical functionalization affects the CNTs rope size and results in exfoliation into smaller bundles and individual nanotubes. These issues are especially important for the fabrication of high-performance composite materials. On the other hand, CNTs are considered to be the ideal reinforcing fibers due to their exceptional mechanical, electronic and thermal properties, low density and high aspect ratio. However, the incorporation of CNTs into the polymer matrix - is often problematic due to the chemical inertness of CNTs. The covalent functionalization of CNTs is a valuable route towards the development of high-performance composites. It provides homogeneously dispersed CNTs incorporated in the polymer and a strong interfacial bonding between the polymer and CNTs. We have applied chemically functionalized SWNTs-COOH to prepare a number of SWNT-polymer composite materials. CNTs can be divided into two main categories: single-walled carbon nanotubes (SWNTs) and multi-walled carbon nanotubes (MWNTs). The first are formed by a single graphene sheet. The latter are formed by additional graphene sheets wrapped around the SWNT core. As a result of the first report concerning the preparation of a CNTs/polymer composite by Ajayan et al., many efforts have been made to combine CNTs and polymers to produce functional composite materials with superior properties. Apart from possible improvements in the mechanical and electrical properties of polymers, the formation of CNTs/polymer composites has been and still is explored as a promising approach for an effective incorporation of CNTs into practical devices [18] -[26] .

In this study is described electrochemical synthesis of nanocomposite films from conducting polymers and functionalized single-walled carbon nanotubes (FSWCNTs) with different dopants and electrochemical characterization of these nanocomposites by cyclic voltammetry, electrochemical impedance spectroscopy, FT-IR and scanning electron microscopy (SEM), have been presented in this paper.

2. EXPERIMENTAL

The electrochemical polymerizations were carried out using a conventional three electrodes system. A platinum electrode and a saturated calomel electrode (SCE) were used as counter and reference electrode, respectively. The reference electrode was placed in a separate cell and was connected to the electrolytic cell via a salt bridge that ends as a Luggin capillary in the electrolytic cell. This arrangement helps in reducing the ohmic resistance of the electrochemical system. The working electrode was made from a platinum disk with surface area of 0.5 cm^2. In this study were used single-walled carbon nanotubes functionalized with carboxylic acid (FSWCNTs) were provided by

CarbonSolutions, Inc. (www.carbonsolution.com, Riverside, CA), and 3,4-ethylenedioxy thiophene (EDOT) (99.5% Fluka) was used as supplied. Which were purchased from Sigma-Aldrich with the following characteristics: 80% - 90% carbon basis, D × L 2 - 10 nm × 0.5 - 2 µm, bundle dimensions. The organic carboxyl groups formed on a nanotube surface are localized at the defects in functionalized single-walled and suitable reactive organic groups from other chemical chains are prone to react in this zone. Bidistilled water was used for all sample preparations. All chemicals were of the highest quality commercially available and were used as received. Cyclic voltammetry and electrochemical impedance spectroscopy were used to investigate the electrochemical properties of the composite films. Electrochemical experiments were carried out with an automated model VoltaLab 40 potentiostat/galvanostat with EIS dynamic controlled by a personal computer. All the following potentials reported in this work are against the SCE (saturated calomel electrode). Scanning electron microscopy (SEM) was used to compare the microstructures of the deposited films.

2.1. Preparation of Modified Electrodes

The Pt electrode was carefully polished with aqueous slurries of fine alumina powder 0.05 micron on a polishing cloth until a mirror finish was obtained. After 20 min sonication, the electrodes were immersed in concentrated H_2SO_4, followed by thorough rinsing with water and ethanol. The prepared electrodes were dried and used for modification immediately. Nanocomposite films of CPs/FSWCNTs were prepared by electrochemical polymerization from a solution containing both the functionalized carbon nanotubes (FSWCNTs) and the corresponding monomer (2,3-dioxythiophene). One kind of FSWCNTs was used in this work namely: single wall carbon nanotubes (SWCNTs with carboxylic acids, see Scheme 1). In the first step, the FSWCNTs aqueous suspension (usually 10 mg/L) was prepared via sonication (1 h) and then the synthesis solution was prepared straightaway by dissolving the monomer and the corresponding supporting electrolyte in the FSWCNTs aqueous suspension. The negatively charged FSWCNTs in solution acted as sole supporting electrolyte and dopant for the (3,4-ethylenedioxy thiophene)-EDOT depositions. For this reason FSWCNTs are enwrapped in polymers during the electropolymerization process in the form of counter ions or dopants. Thus, PEDOT-FSWCNTs composite films were prepared, from an aqueous solution containing 0.025 mol/L of 2,3-dioxythiophene + 0.1 M of $LiClO_4$ + 10 mg/L of FSWCNTs by cyclic voltammetry in the potential scanning range of 0 mV to +1250 mV at a scan rate of 10 mV/s and for a cycles number of 10. The pure poly (3,4-ethylenedioxy thiophene) films were prepared from an aqueous solution of 0.025 mol/L 2,3-dioxythiophene + 0.1 M $LiClO_4$, without adding FSWCNTs, also by cyclic voltammetry in the potential range 0 mV to + 1250 mV at a scan rate of 10 mV/s.

2.2. Characterization of the Modified Electrodes

Electrochemical impedance spectroscopy (EIS) and cyclic voltammetry (CV) were used to investigate the electrochemical properties of the composite films. The electrochemical characterization of the PEDOT-FSWCNTs and PEDOT films was carried out in 0.1 M $LiClO_4$ cycling aqueous solutions for comparison and because the dopant anion of the polymeric films is the same with the anion of the cycling solution. The working electrode potential was cycled in the potential range of 0 mV to +1250 mV with a scan rate of 50 mV/s. The impedance measurements were performed using a VoltaLab 40 potentiostat/galvanostat with EIS dynamic in the frequency range of 100 kHz to 1 mHz with an AC wave of 5 mV (peak-to-peak) overlaid on a DC bias potential and the impedance data were obtained at a rate of 10 points per decade change in frequency. All tests have been performed at 25°C under atmospheric oxygen without agitation.

3. RESULTS AND DISCUSSIONS

The PEDOT film was obtained on the platinum substrate in a synthesis solution of 0.025 mol/L 2,3- dioxythiophene + 0.1 M $LiClO_4$ by cyclic voltammetry in the scanning potential range of 0 to +1250 mV at a scan rate of 10 mV/s and for 10 cycles. It is clearly seen from Figure 1 that, the reaction current increases

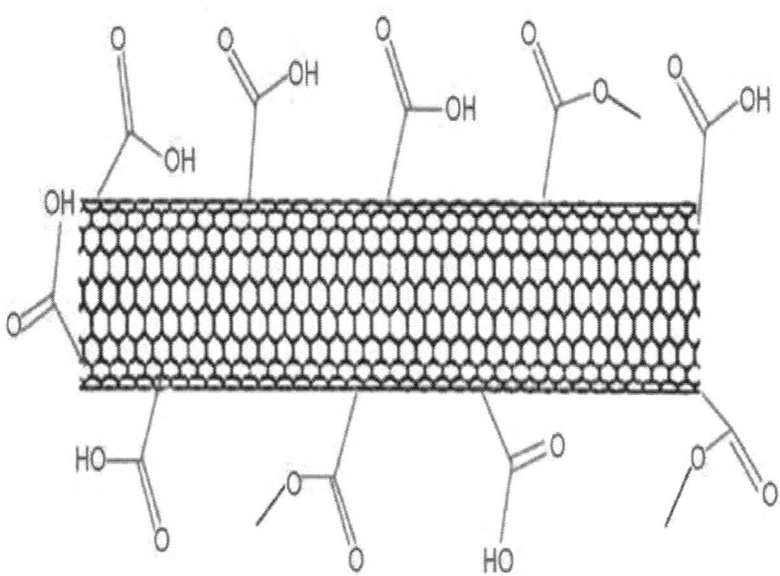

Scheme 1. Chemical structure of single walled carbon nanotube functionalized with carboxylic acids (SWCNTs-COOH).

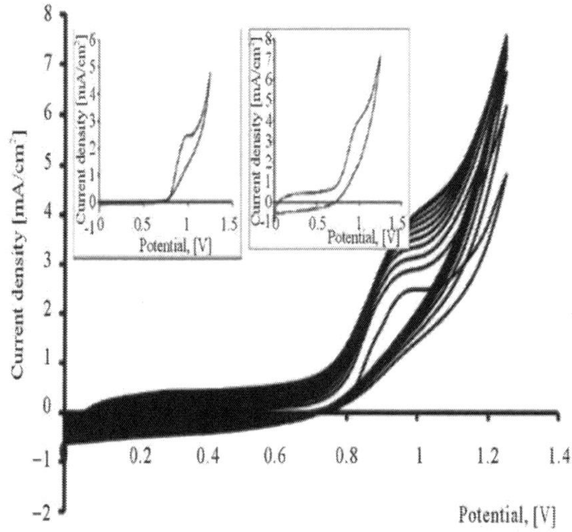

(a)

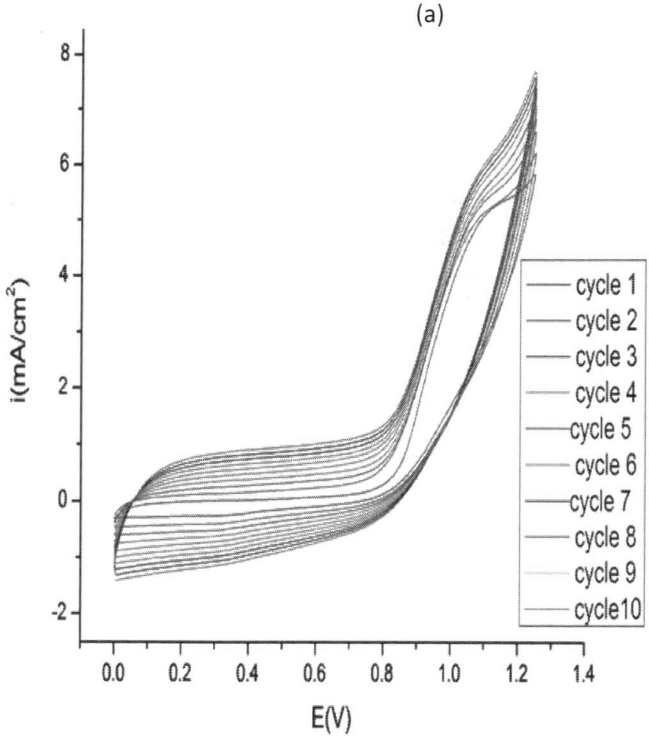

(b)

Figure 1. Polymerization cyclovoltammograms of 2,3-dioxythiophene in 0.1 M $LiClO_4$ aqueous solution. Inset 1st Cycle (a) and 10th Cycle (b).

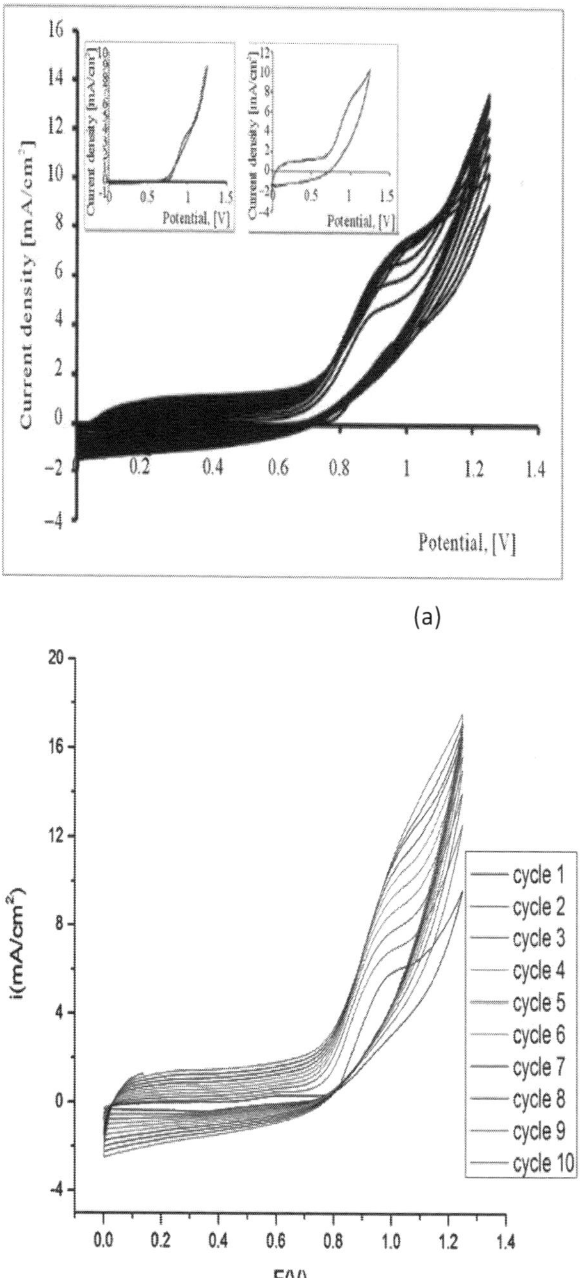

(b)

Figure 2. Polymerization cyclovoltammograms of 2,3-dioxythiophene + CNT-COOH in 0.1 M $LiClO_4$ aqueous. solution. Inset 1st Cycle (a) and 10th Cycle (b).

sharply when the applied potential is larger than 0.7 V, which reveals the formation of the radical cations. The increased current implies that the EDOT radical cations start to electropolymerize onto the platinum electrode. During the first potential scan, the polymerization potential of PEDOT on Pt is around 0.975 V, while the onset of polymerization shifts significantly on following scans to more positive potentials and stabilizes at 1.095 V.

Composite films of poly (3,4-ethylenedioxythiophene) and functionalized, single-walled carbon nanotubes (PEDOT-FSWCNTs) were fabricated by a simple oxidative electropolymerization method. These films were electrodeposited on a platinum working electrode using a classical system with three electrodes. The PEDOT/ CNT-COOH film was obtained on the platinum substrate in a synthesis solution of 0.025 mol/L 2,3-dioxythiophene +10 mg/L carbon nanotubes polycarboxylic acids.

The obtained cyclovoltammograms are given in Figure 2 and have the same shapes with those obtained for PEDOT/Pt modified electrode cycled in the same conditions, but in this case the anodic peaks are much higher and larger than those for PEDOT film. This fact can be explained taking into account that, FSWCNTs (in our case SWCNTs-COOH) are negatively charged and they can act as doped anions and consequently, the conductivity of PEDOT/FSWCNTs composite film increases.

Hence, the PEDOT/FSWCNTs composites exibit higher currents than PEDOT films, which can be translated into larger capacitance. This is an indication of faster kinetics in the composite, which can be attributed to the higher electronic conductivity of the FSWCNTs network. This means that the redox processes which take place in nanocomposite film are more intense and more complex than those in the pure PEDOT film.

The PEDOT/CNT-COOH/dopant films were obtained on the platinum substrate in a synthesis solution of 0.025 mol/L 2,3-dioxythiophene + 10 mg/L CNT-COOH + 0.01 mol/L dopant substance in 0.1 M $LiClO_4$ by cyclic voltammetry in the scanning potential range of 0 to +1250 mV at a scan rate of 10 mV/s and for 10 cycles. The dopant substances used were SDS (Sodium dodecyl sulfate) and tiron (1,2-Dihydroxybenzene-3,5-disulfonic acid disodium salt hydrate). In the presence of dopant (SDS and tiron, see Figure 3), it is observed that the current response of doped PEDOT films decreases with addition of the dopant. Moreover, the current increases with continuous cycling in the potential range as seen in Figure 3. During the first potential scan, the polymerization potential of PEDOT/CNT-COOH/dopant on Pt is around 1.05 V, while the onset of polymerization shifts significantly on following scans to more positive potentials and stabilizes at 1.10 V. All PEDOT composite films were uniform, smooth and adherent to the surface of Pt electrode.

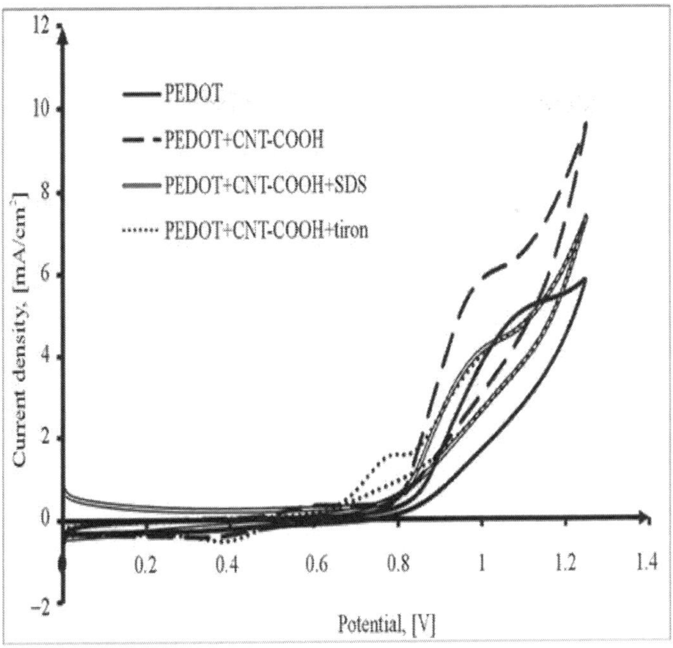

(a)

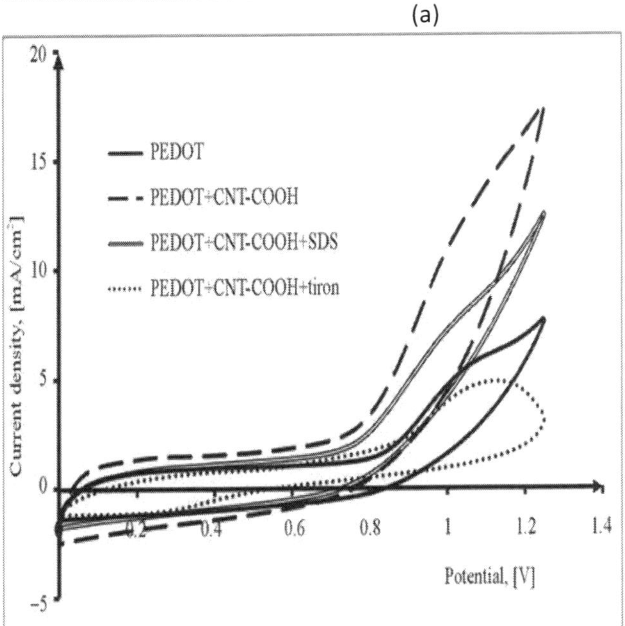

(b)

Figure 3. Comparative polymerization cyclovoltammograms of 2,3-dioxythiophene + CNT-COOH + dopat in 0.1 M $LiClO_4$ aqueous solution, 1st Cycle (a) and 10th Cycle (b).

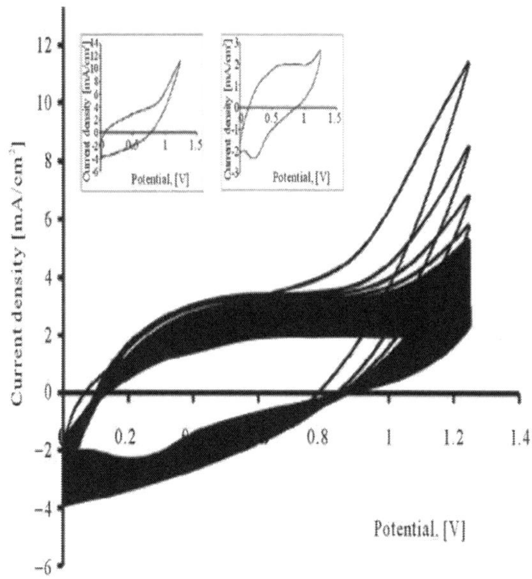

(a)

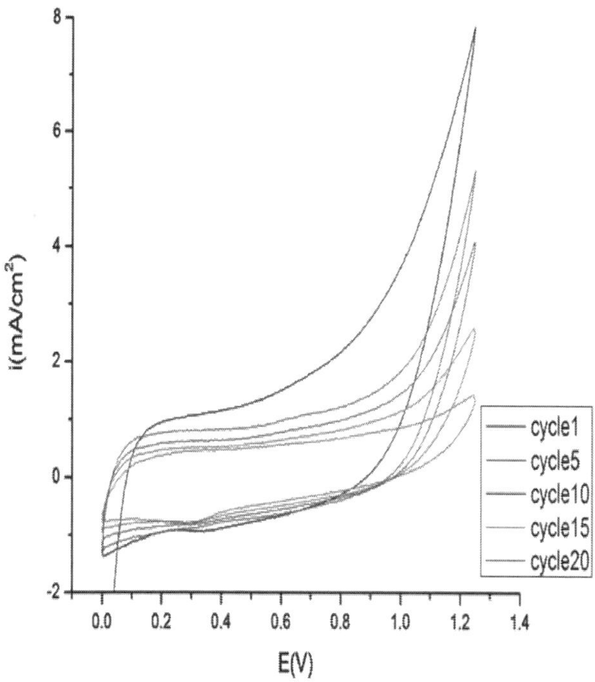

(b)

Figure 4. Cyclic voltammograms of PEDOT film in cycling solution (monomer free) of 0.1 M $LiClO_4$. Inset 1st Cycle (a) and 50th Cycle (b).

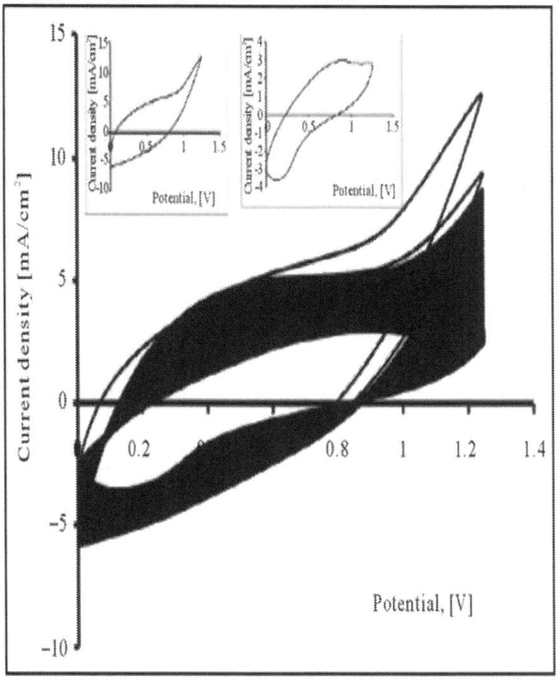

(a)

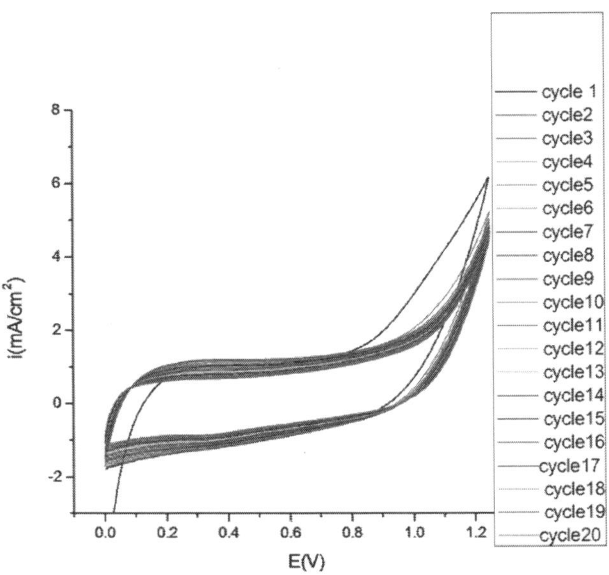

(b)

Figure 5. Cyclic voltammograms of PEDOT/CNT-COOH film in cycling solution (monomer free) of 0.1 M $LiClO_4$. Inset 1st Cycle (a) and 50th Cycle (b).

The electrochemical behavior of the PEDOT, PEDOT/CNT-COOH films deposited electrochemically from aqueous solution in the absence and presence of dopant (SDS and tiron) was characterized further using cyclic voltammetry. The electrochemical characteristics of obtained PEDOT film was study in the cycling solutions, an aqueous solution of 0.1 M $LiClO_4$. The electrode potential was cycled on the potential range from 0 up to 1250 mV with a sweep rate of 50 mV/s and for a cycles number of 20 (see Figure 4).

The electrochemical characteristics of obtained PEDOT/CNT-COOH film was study in the cycling solutions, an aqueous solution of 0.1 M $LiClO_4$. The electrode potential was cycled on the potential range from 0 up to 1250 mV with a sweep rate of 50 mV/s and for a cycles number of 20 (see Figure 5). These films can be cycled repeatedly between the conducting (oxidized) and insulating (neutral) state without significant decomposition of the material, which is consistent with the results reported in the literature [17] .

As can be seen, the curves of these polymer films have nearly rectangular shape, which is typical of the pure capacitive behavior of the tested object [21] -[31] . However, the current of film obtained by potentiodynamic method in presence of dopant is nearly 3 times of the film gained in absence of dopant.

The electrochemical characteristics of obtained PEDOT and PEDOT/CNT-COOH/dopant films were study in the cycling solution, an aqueous solution of 0.1 M $LiClO_4$. The electrode potential was cycled on the potential range from 0 up to 1250 mV with a sweep rate of 50 mV/s and for a cycles number of 20 (see Figure 6).

Analyzing in comparison the obtained results it can be observed that, in all the cases the PEDOT/CNT-COOH composite film reveals current much higher than PEDOT film. This fact can be explained thus, the PEDOT chains became neutral and the negative charge of immobile FCNTSF should be balanced by the cations with small size from the supporting electrolyte solution.

The surface morphology, formation mechanism and electrochemical nature of PEDOT-FSWCNTs films were investigated using scanning electron microscopy (SEM), cyclic voltammetry (CV) and alternating current (AC) impedance spectroscopy (EIS). CV (cyclic voltammetry) and EIS (electrochemical impedance spectroscopy) revealed that the PEDOT-FSWCNTs electrode had higher electrocatalytic activity for the redox reaction and a smaller charge transfer resistance than the PEDOT electrodes.

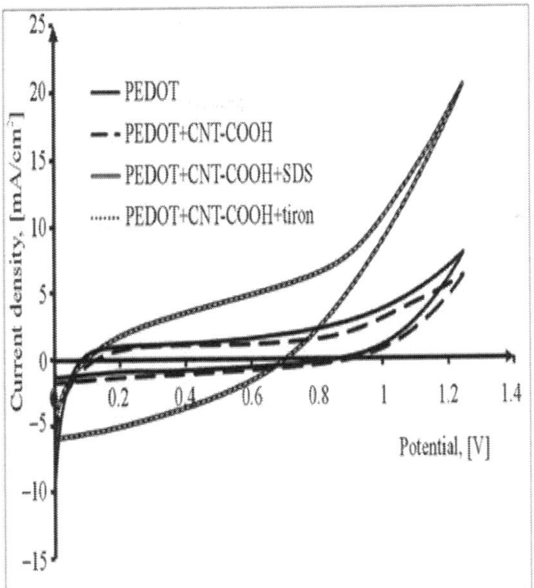

(a)

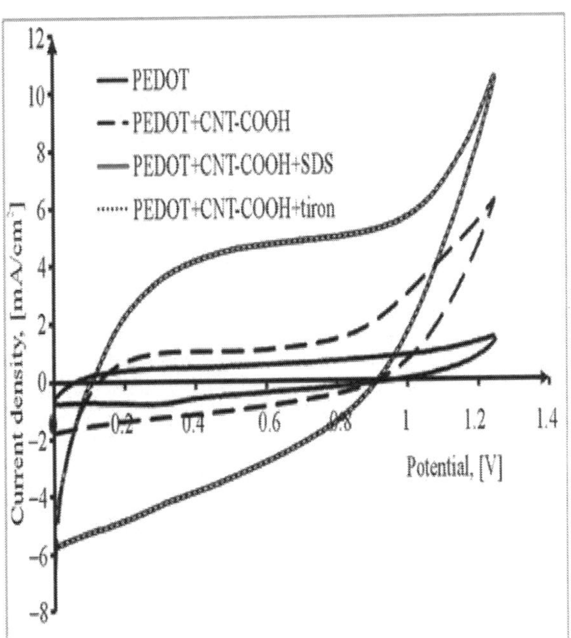

(b)

Figure 6. Comparative cyclic voltammograms of PEDOT, PEDOT/ CNT-COOH with dopants films in cycling solution (monomer free) of 0.1 M $LiClO_4$, 1st Cycle (a) and 20th Cycle (b).

Further, the composite and pure polymer films were studied by EIS at open circuit potential, in an aqueous solution of 0.1 M $LiClO_4$ and 25°C. The resulting Nyquist plots and Bode plots for PEDOT, PEDOT/SWCNTsCOOH and PEDOT/SWCNTs-COOH/dopant systems are shown in Figure 7 and Figure 8. The impedance plot is composed of a semicircle at high frequencies and a capacitive slope at low and middle frequencies (see Figure 7 and Figure 8). The semicircle appeared at high frequencies is considered to owe to the charge transfer resistance, which originates from the interface structure between the porous electrode surface and the electrolyte [19] . At low frequencies, the impedance plot becomes a near vertical line. The Nyquist plots for both PEDOT, PEDOT/SWCNTs-COOH and PEDOT/SWCNTs-COOH/dopant composite films are featured by a vertical trend at low frequencies, indicating a capacitive behaviour according to the equivalent circuit theory. Bode diagrams point out also the capacitive behaviour in concordance with Nyquist plots (see in comparison Figures 7(a)-(d) and Figures 8(a)-(d)).

The capacitances of the electrode materials were calculated, according to the equation:

$$C = -1/(2\pi f Z_{im}).$$

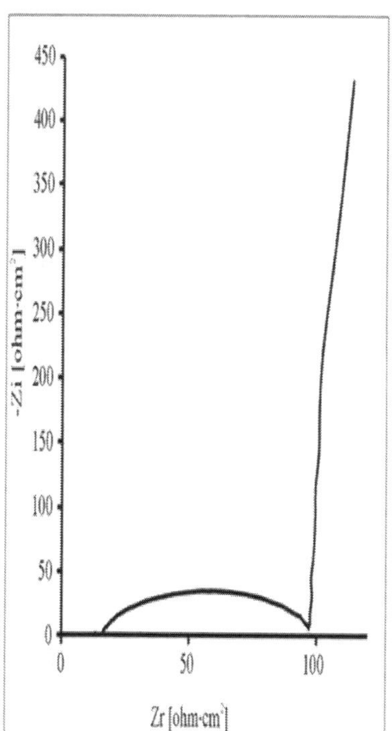

(a)

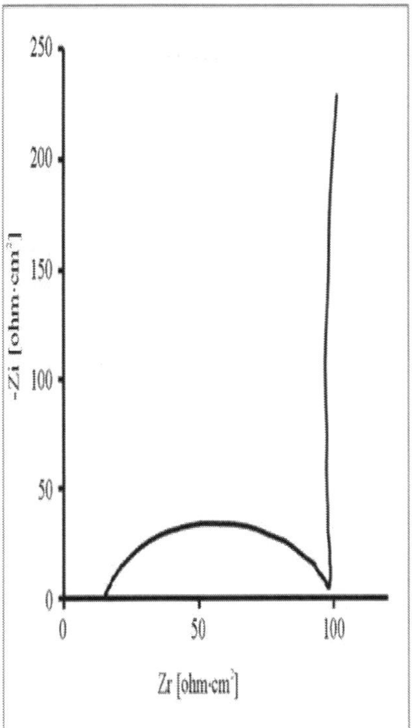
250
200
150
100
50
0
-Zi [ohm·cm²]
0
50
100
Zr [ohm·cm²]

(b)

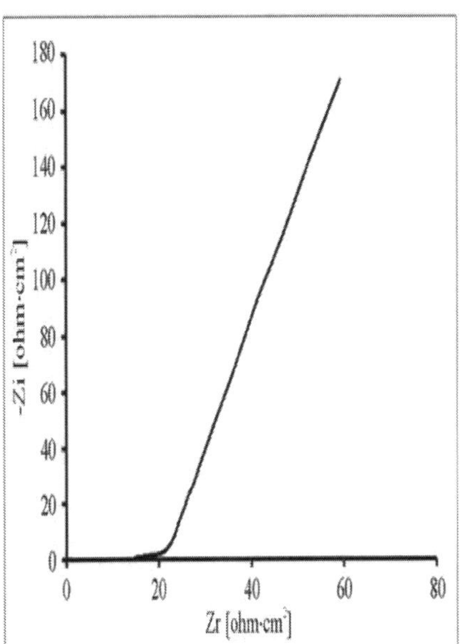
180
160
140
120
100
80
60
40
20
0
-Zi [ohm·cm²]
0
20
40
60
80
Zr [ohm·cm²]

(c)

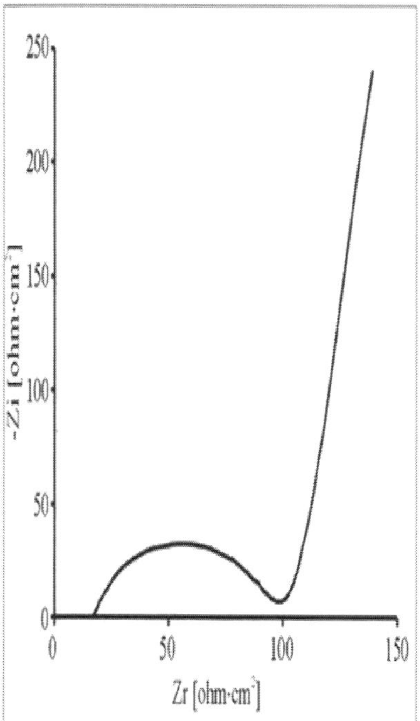

(d)

Figure 7. The Nyquist diagrams for modified electrodes (a) PEDOT/Pt, (b) PEDOT/CNsT-COOH/Pt, (c) PEDOT/CNTsCOOH/SDS/Pt and (d) PEDOT/CNTs-COOH/tiron/Pt at open circuit potential in an aqueous solution of 0.1 M $LiClO_4$.

(f = frequency; Z_{im} = imaginary impedance), from the slope of the linear correlation between the imaginary impedance and the reciprocal of the frequency at low frequencies.

From these Figures 7-9 and Table 1, one can observe higher capacitance the one order of magnitude value for PEDOT/FSWCNTs + dopant film in respect with PEDOT pure polymeric films. Higher capacitance of the composite films results obviously from the contribution of the embedded FSWCNTs that provide interconnected pathways for electrons through the FSWCNTs and ions through the pore network or the direct interaction between the delocalised electrons on polymer chains and the FSWCNTs.

The real impedance at low frequencies, where the capacitive behaviour dominates, is an indication of the combined resistance of the electrolyte and the film including both electronic and ionic contributions. The values of the real impedance at 0.01 Hz are given also in Table 1. It can be seen that the PEDOT/FSWCNTs films were significantly lower in resistance than PEDOT films.

It can also be seen that PEDOT/FSWCNTs offered much higher overall conductivity compared with the PEDOT film. It has been already mentioned that, in general, the real impedance of an electrode material also decreases as the material's porosity increases due to improved ionic accessibility [20] -[27] [32] -[41] .

Figure 10 shows the SEM images of the PEDOT film. In Figure 10(a), a nodular accumulating structure is evidenced. The size of the nodules ranged a few hundred nanometers in diameter and they aggregate to form gobbets. In Figure 10(b), the effect of dopant on the morphology of PEDOT film can be observed: lamellar

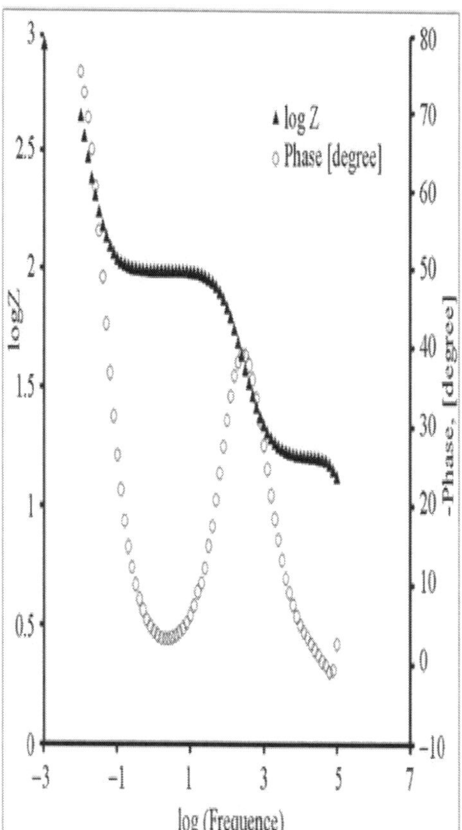

(a)

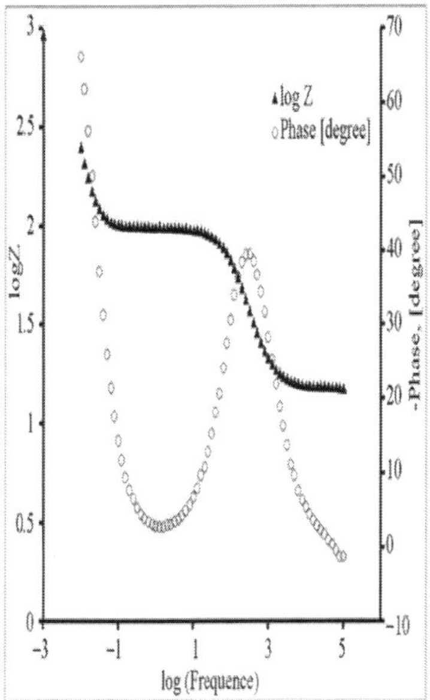
log Z
Phase [degree]
logZ
-Phase, [degree]
log (Frequence)

(b)

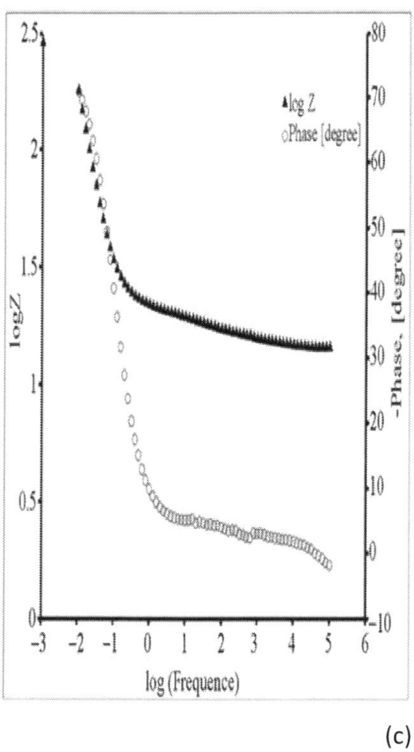
log Z
Phase [degree]
logZ
-Phase, [degree]
log (Frequence)

(c)

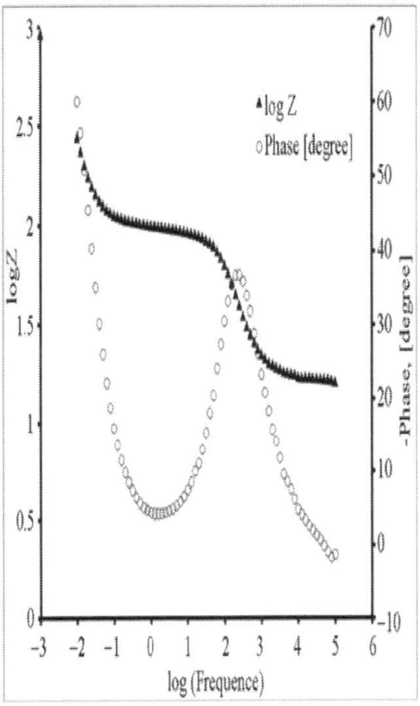

(d)

Figure 8. The Bode diagrams for modified electrodes (a) PEDOT/Pt, (b) PEDOT/CNTs-COOH/Pt, (c) PEDOT/CNTsCOOH/SDS/Pt and (d) PEDOT/CNTs-COOH/tiron/Pt at open circuit potential in an aqueous solution of 0.1 M $LiClO_4$.

Table 1. Real impedance and capacitance values of pure PEDOT film and PEDO/FSWCNTs nanocomposite film with different dopants obtained by co-polymerization using the cyclic voltammetry (CV) at 0.01 Hz.

Polymeric film	Slope values obtained from graph $-Z'' = f(1/2\pi f)$	C [F/cm^2]	Z_r (Ωcm^2) at 0.01 Hz
PEDOT	103	0.009	357
PEDOT/CNT-COOH	72	0.014	267
PEDOT/CNT-COOH/SDS	15	0.067	120
PEDOT/CNT-COOH/tiron	11	0.091	139

structure with almost vertical orientation to the substrate. It is evident that dopant changes the morphology of PEDOT film into a more porous structure with higher interface area. This is in agreement with the SEM results presented below that illustrated a smaller porosity in the PEDOT film than in the case of the composite film with dopant. The SEM images showed that nanocomposite films were more porous than PEDOT films.

Fourier transform infrared (FT-IR see Figure 11) spectra were carried with a Bruker optics spectrometer at room temperature. All spectra in this paper were obtained at a resolution 4 cm^{-1} in the region 4000 - 600 cm^{-1}.

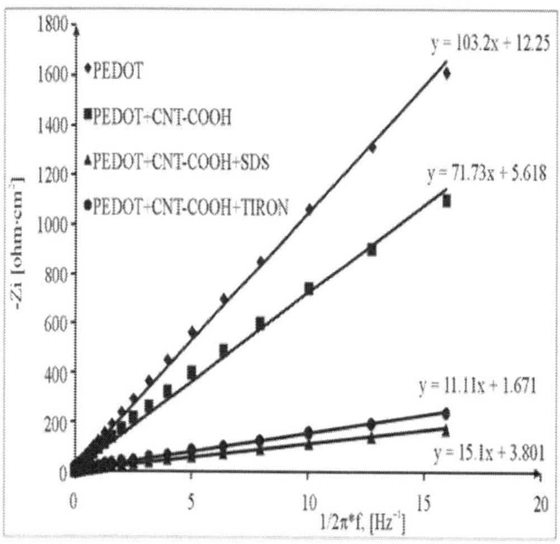

Figure 9. Capacitance evaluation for PEDOT/Pt, PEDOT/CNT/Pt, PEDOT/ CNT-COOH/SDS/Pt and PEDOT/CNT-COOH/ tiron/Pt modified electrode.

(a)

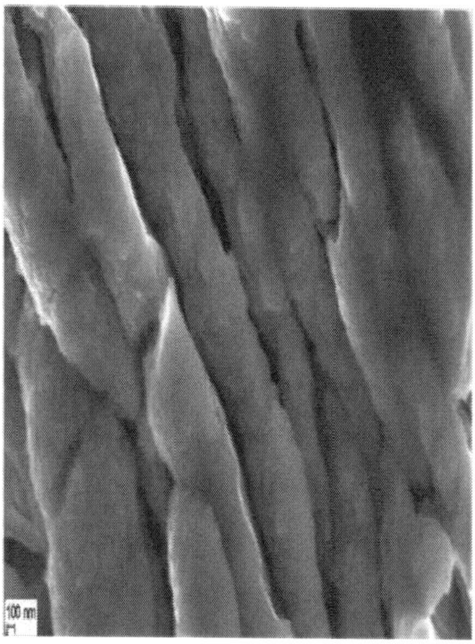

(b)

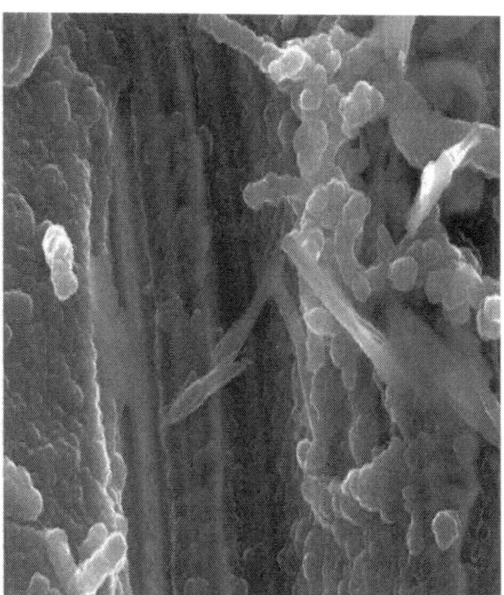

(c)

Figure 10. SEM images of the film surface of (a) PEDOT and (b) PEDOT doped film formation by cyclic voltammetry (0 to 1250 mV at a scan rate of 10 mV/s), (c) PEDOT/CNT-COOH.

FT-IR spectrometer is a powerful instrument that can be used to determine type of bonding for to obtain a new composite. The characteristic bands in the FT-IR spectrum for PEDOT and nanocomposite are the following: a very weak and a medium band at 3000 - 4000 cm^{-1} is assigned to the C-H and OH stretching modes; vibration at 1521, 1476, 1328 and 922 cm^{-1} for PEDOT+ nanocompozit and 1501, 1403, 1181 and 916 cm^{-1} for PEDOT are

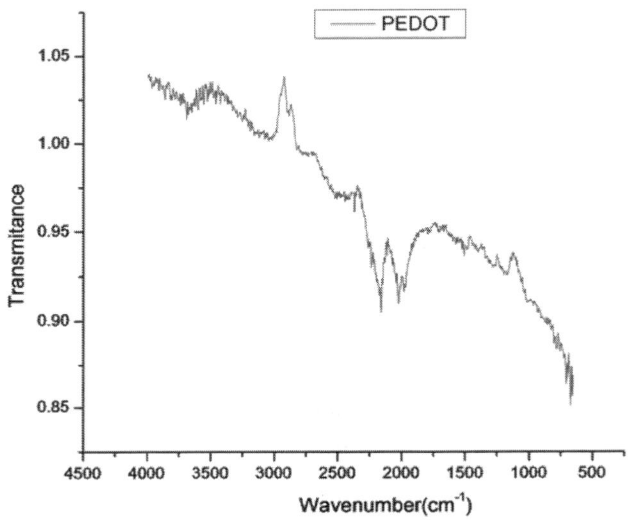

(a)

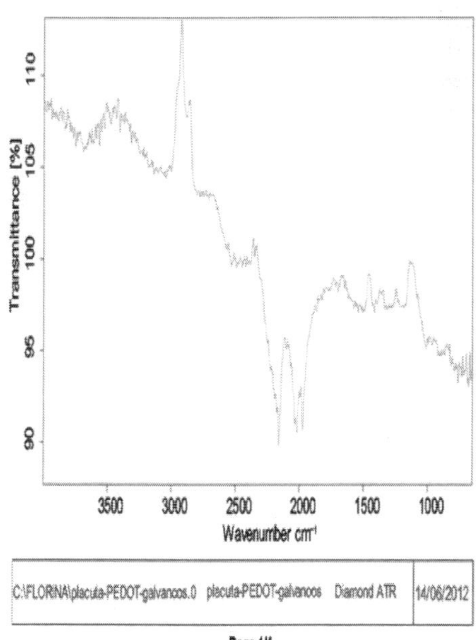

(b)

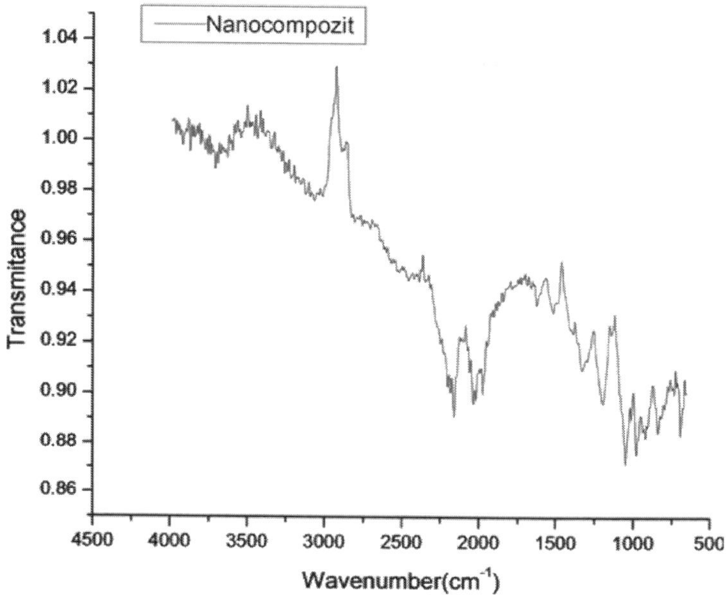

(c)

Figure 11. FT-IR spectra of PEDOT and nanocompozit.

attributed to the stretching modes of C=C, C-C, and C-S in the thiophene ring [40] [42] -[44] . The bands at 1203 and 1046 cm^{-1} are assigned to the stretching modes of the ethylenedioxy (alkylenedioxy) group and the band around 1046 and 922 cm^{-1} is due to the ethylenedioxy ring deformation mode. The spectrum shows several bands of which the band at 822 cm^{-1} for nanocompozit and 815 cm^{-1} for PEDOT is assigned to the symmetric C-S-C deformation. The band at 1046 cm^{-1} (nanocompozit) and 1412 cm^{-1} (PEDOT) is assigned to the symmetric C-O-C ether bond. A difference that can be observed is the intensity ratio of bands the spectrum (1500 - 600 cm^{-1}) of nanocompozit film than PEDOT band, exhibits a clear difference in intensity [45] . This may suggest that the dopant promotes and stabilizes the structure of the film (nanocomposite).

4. CONCLUSIONS

Electrochemically synthesized composite films of conducting polymers (PEDOT) and FSWCNTs with different dopants have in common a porous structure at nano-meter scales. They have better mechanical integrity and a higher conductivity than the similarly prepared pure polymer. The electrochemical activity of PEDOT/FSWCNTs/Pt modified electrode in 0.1 M $LiClO_4$ cycling solution is much more higher than of PANI/Pt modified electrode in the same cycling solution.

The Nyquist plots for both PEDOT and composite films are featured by a vertical trend at low frequencies, indicating a capacitive behaviour according to the equivalent circuit theory. The higher capacitance of the composite films results obviously from the contribution of the embedded FSWCNTs that provide interconnected pathways for electrons through the FSWCNTs and ions through the pore network or the direct interaction between the delocalised electrons on polymer chains and the FSWCNTs.

The PEDOT and FSWCNTsF with dopants films reveal a morphological structure composed of nanofibers, but for the pure polymeric ones they are slightly thicker than those in the PANI/FSWCNTs composite film.

Microstructures of these composites suggests that PEDOT was wrapped around FSWCNTs. The obtained results can be employed to get the desired value of the capacitance by choosing the adequate preparation method and so, by controlling the microstructure of the composites.

ACKNOWLEDGEMENTS

Financial support from PN-II-ID-PCE-2008-2 contract number 596, code ID_716 (The National University Research Council) is gratefully acknowledged.

REFERENCES

1. Ajayan, P.M., Stephan, O., Colliex, C. and Trauth, D. (1994) Aligned Carbon Nanotube Arrays Formed by Cutting a Polymer Resin—Nanotube Composite. Science, 265, 1212-1214.
2. Dai, L. and Mau, A.W.H. (2001) Controlled Synthesis and Modification of Carbon Nanotubes and C_{60}: Carbon Nanostructures for Advanced Polymeric Composite Materials. Advanced Materials, 13, 899-913.
3. Baughman, R.H., Zakhidov, A.A. and Heer, W.A. (2002) Carbon Nanotubes—the Route toward Applications. Science, 297, 787-792.
4. Liao, Y.Z., Zhang, C., Zhang, Y., Strong, V., Tang, J.S., Li, X.-G., Kalantar-Zadeh, K., Hoek, E., Wang, K. and Kaner, R. (2011) Carbon Nanotube/Polyaniline Composite Nanofibers: Facile Synthesis and Chemosensors. Nano Letters, 11, 954-959.
5. Dai, L. (1999) Advanced Syntheses and Microfabrications of Conjugated Polymers, C_{60}-Containing Polymers and Carbon Nanotubes

for Optoelectronic Applications. Polymers for Advanced Technologies, 10, 357-420.

6. Chen, G.Z., Shaffer, M.S.P., Coleby, D., Dioxan, G., Fray, D.J. andWindle, A.H. (2000) Carbon Nanotube and Polypyrrole Composites: Coating and Doping. Advanced Materials, 12, 522-526.

7. Saito, Y., Uemura, S. and Hamaguchi, K. (1998) Cathode Ray Tube Lighting Elements with Carbon Nanotube Field Emitters. Japanese Journal of Applied Physics, 37, L346-L348.

8. Liao, Y., Zhang, C., Wang, X., Li, X.G., Ippolito, S.J., Kalantar-Zadeh, K. and Kaner, R.B. (2011) Carrier Mobility of Single-Walled Carbon Nanotube-Reinforced Polyaniline Nanofibers. The Journal of Physical Chemistry C, 115, 16187-16192.

9. Prună, A., Branzoi, V. and Branzoi, F. (2011) Ordered Arrays of Copper Nanowires Enveloped in Polyaniline Nanotubes. Journal Applied Electrochemistry, 41, 77-81.

10. Shirakwa, H. (2001) The Discovery of Polyacetylene Film: The Dawning of an Era of Conducting Polymers (Nobel Lecture). Angewandte Chemie International Edition, 40, 2574-2580.

11. Groenendaal, B.L., Jonas, F., Freitag, D., Pielartzik, H. and Reynolds, J.R. (2000) Poly(3,4-Ethylenedioxythiophene) and Its Derivatives: Past, Present, and Future. Advanced Materials, 12, 481-494.

12. Crispin, X., Jakobsson, F.L.E., Crispin, A., Grim, P.C.M., Andersson, P., Volodin, A., van Haesendonck, C., Van der Auweraer, M., Salaneck, W.R. and Berggren, M. (2006) The Origin of the High Conductivity of Poly(3,4-Ethylenedioxythiophene)-Poly(Styrenesulfonate) (PEDOT-PSS) Plastic Electrodes. Chemistry of Materials, 18, 4354-4360.

13. Breiby, D.W., Samuelsen, L.B. and Groenendaal, B. (2003) Smectic Structures in Electrochemically Prepared Poly(3,4-Ethylenedioxythiophene) Films. Journal of Polymer Science Part B: Polymer Physics, 41, 945-952.

14. Winther-Jensen, B. and West K. (2006) Stability of Highly Conductive Poly-3,4-Ethylene-Dioxythiophene. Reactive and Functional Polymers, 66, 479-483.

15. Kros, A., Sommerdijk, N.A.J.M. and Nolte R.J.M. (2005) Poly(Pyrrole) versus Poly(3,4-Ethylenedioxythiophene): Implications for Biosensor Applications. Sensors and Actuators B: Chemical, 106, 289-295.

16. Luo, S.C., Mohamed Ali, E., Tansil, N.C., Yu, H.H., Gao, S. and Ying, J.Y. (2008) Poly(3,4-ethylenedioxythiophene) (PEDOT) Nanobiointerfaces:

Thin, Ultrasmooth, and Functionalized PEDOT Films with in Vitro and in Vivo Biocompatibility. Langmuir, 24, 8071-8077.

17. Du, X. and Wang, Z. (2003) Effects of Polymerization Potential on the Properties of Electrosynthesized PEDOT Films. Electrochimica Acta, 48, 1713-1717.

18. Li, L., Loveday, D.C., Mudigonda, D.S.K. and Ferraris, J.P. (2002) Effect of Electrolytes on Performance of Electrochemical Capacitors Based on Poly[3-(3,4-difluorophenyl)thiophene]. Journal of the Electrochemical Society, 149, A1201-A1207.

19. Branzoi, V., Prună, A. and Branzoi, F. (2008) Electrosynthesized Doped Polypyrrole Films on Pure Zinc Electrode. Molecular Crystal and Liquid Crystals, 485, 853-861.

20. Branzoi, V., Branzoi, F. and Pilan, L. (2009) Characterization of Electrodeposited Polymeric and Composite Modified Electrodes on Cobalt Based Alloy. Materials Chemistry and Physics, 118, 197-203.

21. Ago, H., Petritch, K., Shaffer, M.S.P., Windle, A.H. and Friend, R.H. (1999) Composites of Carbon Nanotubes and Conjugated Polymers for Photovoltaic Devices. Advanced Materials, 11, 1281-1285.

22. Hughes, M., Chen, G.Z., Shaffer, M.S., Fray, D.J. and Windle, A.H. (2002) Electrochemical Capacitance of a Nanoporous Composite of Carbon Nanotubes and Polypyrrole. Chemistry of Materials, 14, 1610-1613.

23. Downs, C., Nuget, J., Ajayan, P.M., Duquette, D.J. and Santhanam, K.S.V. (1999) Growth from Surface Methodology for the Fabrication of Functional Dual Phase Conducting Polymer Polypyrrole/Polycarbazole/Polythiophene (cp/Polypyr/Polycbz/Polyth)-Carbon Nanotube (cnt) Composites of Controlled Morphology and Composition—Sidewall Vers. Advanced Materials, 12, 1028.

24. Peng, C., Jin, J. and Chen, G.Z. (2007) A Comparative Study on Electrochemical Co-Deposition and Capacitance of Composite Films of Conducting Polymers and Carbon Nanotubes. Electrochimica Acta, 53, 525-537.

25. Wang, J., Dai, J. and Yarlagadda, T. (2005) Carbon Nanotube-Conducting-Polymer Composite Nanowires. Langmuir, 21, 9-12.

26. Tsai, Y.C., Li, S.C. and Liao, S.W. (2006) Electrodeposition of Polypyrrole-Multiwalled Carbon Nanotube-Glucose Oxidase Nanobiocomposite Film for the Detection of Glucose. Biosensors and Bioelectronics, 22, 495-500.

27. Chen, J., Hamon, M.A., Hu, H., Chen, Y., Rao, A.M., Eklund, P.C. and Haddon, R.C. (1998) Solution Properties of Single-Walled Carbon Nanotubes. Science, 282, 95-98.

28. Chen, Y., Haddon, R.C., Fang, S., Rao, A.M., Eklund, P.C., Lee, W.H., Dickey, E.C., Grulke, E.A., Pendergrass, J.C., Chavan, A., Haley, B.E. and Smalley, R.E. (1998) Chemical Attachment of Organic Functional Groups to SingleWalled Carbon Nanotube Material. Journal of Materials Research, 13, 2423-2431.

29. Holzinger, M., Vostrowsky, O., Hirsch, A., Hennrich, F., Kappes, M., Weiss, R. and Jellen, F. (2001) Sidewall Functionalization of Carbon Nanotubes. Angewandte Chemie International Edition, 40, 4002-4005.

30. Dyke, C.A. and Tour, J.M. (2003) Unbundled and Highly Functionalized Carbon Nanotubes from Aqueous Reactions. Nano Letters, 3, 1215-1218.

31. Saini, R.K., Chiang, I.W., Peng, H., Smalley, R.E., Billups, W.E., Hauge, R.H. and Margrave, J.L. (2003) Covalent Sidewall Functionalization of Single Wall Carbon Nanotubes. Journal of the American Chemical Society, 125, 3617- 3621.

32. Dyke, C.A. and Tour, J.M. (2003) Solvent-Free Functionalization of Carbon Nanotubes. Journal of the American Chemical Society, 125, 1156-1157.

33. Williams, K.A., Veenhuizen, P.T.M., de la Torre, B.G., Eritja, R. and Dekker, C. (2002) Nanotechnology: Carbon Nanotubes with DNA Recognition. Nature, 420, 761.

34. Niyogi, S., Hamon, M.A., Hu, H., Zhao, B., Bhowmik, P., Sen, R., Itkis, M.E. and Haddon, R.C. (2002) Chemistry of Single-Walled Carbon Nanotubes. Accounts of Chemical Research, 35, 1105-1113.

35. Kamaras, K., Itkis, M.E., Hu, H., Zhao, B. and Haddon, R.C. (2003) Covalent Bond Formation to a Carbon Nanotube Metal. Science, 301, 1501.

36. Chen, J., Rao, A.M., Lyuksyutov, S., Itkis, M.E., Hamon, M.A., Hu, H., Cohn, R.W., Eklund, P.W., Colbert, D.T., Smalley, R.E. and Haddon, R.C. (2001) Dissolution of Full-Length Single-Walled Carbon Nanotubes. Journal of Physical Chemistry B, 105, 2525-2528.

37. Zhao, B., Hu, H. and Haddon, R.C. (2004) Synthesis and Properties of a Water-Soluble Single-Walled Carbon Nanotube-Poly(m-aminobenzene sulfonic acid) Graft Copolymer. Advanced Functional Materials, 14, 71-76.

38. Jurewiez, K., Delpeux, S., Bertagna, V., Beguin, F. and Frackowiak, E. (2001) Supercapacitors from Nanotubes/Polypyrrole Composites. Chemical Physics Letters, 347, 36-40. http://dx.doi.org/10.1016/S0009-2614(01)01037-5

39. Gupta, V. and Miura, N. (2006) Polyaniline/Single-Wall Carbon Nanotube (PANI/SWCNT) Composites for High Performance Supercapacitors. Electrochimica Acta, 52, 1721-1726.

40. Xiao, Q. and Zhou, X. (2003) The Study of Multiwalled Carbon Nanotube Deposited with Conducting Polymer for Supercapacitor. Electrochimica Acta, 48, 575-580.

41. Zeng, J.X., Wei, W.Z., Wu, L., Liu, X.Y., Liu, K. and Li, Y. (2006) Fabrication of Poly(toluidine blue O)/Carbon Nanotube Composite Nanowires and Its Stable Low-Potential Detection of NADH. Journal of Electroanalytical Chemistry, 595, 152-160.

42. Ferrer-Anglada, N., Kaempgen, M. and Roth, S. (2006) Transparent and Flexible Carbon Nanotube/Polypyrrole and Carbon Nanotube/Polyaniline pH Sensors. Physica Status Solidi (b), 243, 3519-3523.

43. Santhosh, P., Manesh, K.M., Lee, K.P. and Gopalan, A.I. (2006) Enhanced Electrocatalysis for the Reduction of Hydrogen Peroxide at New Multiwall Carbon Nanotube Grafted Polydiphenylamine Modified Electrode. Electroanalysis, 18, 894-903.

44. Lu, G.W., Li, C., Shen, J.Y., Chen, Z.J. and Shi, G.Q. (2007) Preparation of Highly Conductive Gold-Poly(3,4-ethylenedioxythiophene) Nanocables and Their Conversion to Poly(3,4-ethylenedioxythiophene) Nanotubes. Journal of Physical Chemistry C, 111, 5926-5931.

45. Ingan, O. (1997) Handbook of Conducting Polymers. Vol. 2, Wiley, New York, 785.

Chapter 3

Aflatoxin B1 Detection Using a Highly-Sensitive Molecularly-Imprinted Electrochemical Sensor Based on an Electropolymerized Metal Organic Framework

Mengjuan Jiang [2], Mohamed Braiek [1], Anca Florea [1], Amani Chrouda [1], Carole Farre [1], Anne Bonhomme [1], Francois Bessueille [1], Francis Vocanson [3], Aidong Zhang [2] and Nicole Jaffrezic-Renault [1],

[1]*University of Lyon, Institute of Analytical Sciences, UMR-CNRS 5280, La Doua Street, 5, Villeurbanne 69100, France*

[2] *Key Laboratory of Pesticide and Chemical Biology of Ministry of Education, College of Chemistry, Central China Normal University, Wuhan 430079, China*

[3]*University of Lyon, Laboratoire Hubert Curien, UMR 5516, Jean-Monnet University of Saint-Etienne, Saint-Etienne F-42023, France*

ABSTRACT

A sensitive electrochemical molecularly-imprinted sensor was developed for the detection of aflatoxin B1 (AFB1), by electropolymerization of p-aminothiophenol-functionalized gold nanoparticles in the presence of AFB1 as a template molecule. The extraction of the template leads to the formation of cavities that are able to specifically recognize and bind AFB1 through π-π interactions between AFB1 molecules and aniline moities. The performance of the developed sensor for the detection of AFB1 was investigated by linear sweep voltammetry using a hexacyanoferrate/hexacyanoferrite solution as a redox probe, the electron transfer rate increasing when the concentration of AFB1 increases, due to a p-doping effect. The molecularly-imprinted sensor exhibits a broad linear range, between 3.2 fM and 3.2 μM, and a quantification limit of 3 fM. Compared to the non-imprinted sensor, the imprinting factor was found to be 10. Selectivity studies were also performed towards the binding of other aflatoxins and ochratoxin A, proving good selectivity.

Keywords: electrochemical sensors; aflatoxin B1; molecularly imprinted polymers; metal organic framework; gold nanoparticles

1. INTRODUCTION

Aflatoxin B1 (Figure 1) (CAS No. 1162-65-8) whose formal name is 2,3,6aR,9aS-tetrahydro-4-methoxy-1H,11H-cyclopenta[c]furo[3',2':4,5]furo[2,3-h][1]benzopyran-1,11-dione is an aflatoxin produced by Aspergillus flavus and A. parasiticus. It is arguably the most potent carcinogen known, and is up to twice as carcinogenic as an equitoxic dose of X-rays. Aflatoxin B1 can permeate through the skin. Dermal exposure to this aflatoxin, in particular environmental conditions, can lead to major health risks. AFB1 has a strong toxicity to the human and some animals and it is the main cause of liver damage. The minimum admissible level of AFB1 in food products should be 20 μg/kg [1,2].

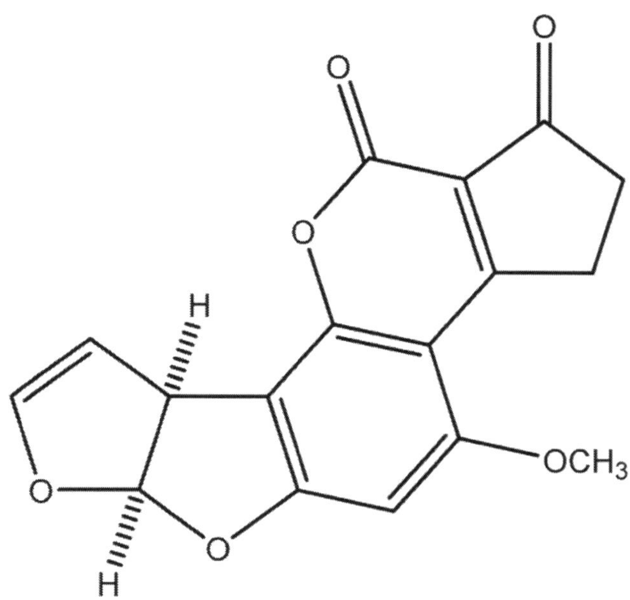

Figure 1. Chemical structure of aflatoxin B_1.

Therefore, determining aflatoxin B1 mainly in almonds, Brazil nuts, hazelnuts and pistachios, and dried figs and other food is very important. Several methods have been reported for the quantification of aflatoxin B1 in food, such as thin layer chromatography (TLC) [3], liquid chromatography (LC) [4], enzyme linked immunosorbent assay (ELISA) [5], liquid chromatography-tandem mass spectrometry (LC-MSMS) [6]. However, these methods suffer from being time consuming, require expensive instrumentation, complex pretreatment of samples, and skilled staff. Increasing efforts have been focused on using electrochemical sensors to measure the concentration of mycotoxins because of their short response time, convenience, high sensitivity, and low cost.

The molecular imprinting technique has been extensively applied in preparation of polymer materials able to recognize small molecules in recent years [7,8,9]. The method lies in the polymerization and crosslinking of functional monomers in the presence of a template molecule, which is subsequently extracted from the resulting polymer matrix, generating the cavities complementary in shape and sizes with the template. The material is obtained with the advantage of high affinity and specificity, good physical and chemical stability, ease of preparation, low cost, and possibility of use in harsh environmental conditions. Molecularly-imprinted polymers (MIPs) found potential applications in many fields. The target molecular recognition ability of MIPs can be assessed by the electrochemical, optical or piezoelectric methods [10,11,12,13]. Immunosensors with molecularly-imprinted polymers are bound to be an alternative to traditional immunosensors based on antibodies. This is due to the unique combination of advantages displayed by the artificial materials, including the absence of animal inoculation and sacrifice, unnecessary hapten conjugation to carrier protein for stimulated production, the possibility of manufacturing MIPs against toxic substances, excellent physicochemical stability, reusability, ease of storage, and recognition in organic media [14].

The preparation of MIPs by direct electropolymerization on the sensing surface eliminates the necessity of strict synthesis and film that usually requires preparing spin- or solution-casting technology, offering the advantage of better control of the film thickness at the electrode surface [15]. Integration of nanoparticles in molecularly-imprinting sensors has the benefit of enhancing the catalytic activity of the modified electrode surface [8].

First defined by Yaghi and co-workers in 1995 [16,17], following the pioneering work of Hoskins and Robson, metal-organic frameworks (MOFs) are crystalline porous hybrid materials comprising coupling units (metal ions or metal-oxo units) coordinated by electron-donating organic ligands. MOFs have the highest specific surface areas reported to date for porous materials (up to 10^4 m^2 g^{-1}) and large internal pore volumes with well-defined pore sizes and apertures. Their principal applications are gas storage, electrode materials for batteries, electrocatalysts for important reactions taking place in fuel cells or electrolyzers, electrode materials for supercapacitors, Li^+ or H^+ conductive materials for batteries or fuel cells, and surface films for corrosion inhibition [18,19,20]. Metal organic frameworks (MOF) as highly selective and selective platforms for the development of sensors have attracted growing interests, associated with a fluorimetric detection [21] or with an electrochemical detection [22,23].

Herein, an electrochemical sensor for the sensitive and selective detection of aflatoxin B1, based on a molecularly-imprinted MOF with recognition sites for aflatoxin B1, is prepared through electropolymerization of p-aminothiophenol (PATP)-functionalized AuNPS in the presence of the AFB1 as a template molecule, combining the advantages of molecular imprinting and electrodeposition with those conferred by MOF. The schematic principle of the fabrication of this MIP-MOF sensor for AFB1 detection is presented in Figure 2. To the best of our knowledge this is the first molecularly-imprinted MOF electrochemical sensor reported to detect aflatoxin B1.

Figure 2. Schematic illustration of the fabrication of the MIP sensor for aflatoxin B1 detection.

2. RESULTS AND DISCUSSION

2.1. Fabrication of Aflatoxin B1 Imprinted MOF Film

The fabrication of the MIP sensor was carried out in several steps as shown in Figure 2.

2.1.1. Characterization of PATP Functionalized AuNPs

TEM observation presented in Figure 3, depicts a mean diameter of around 5 nm with a narrow size distribution.

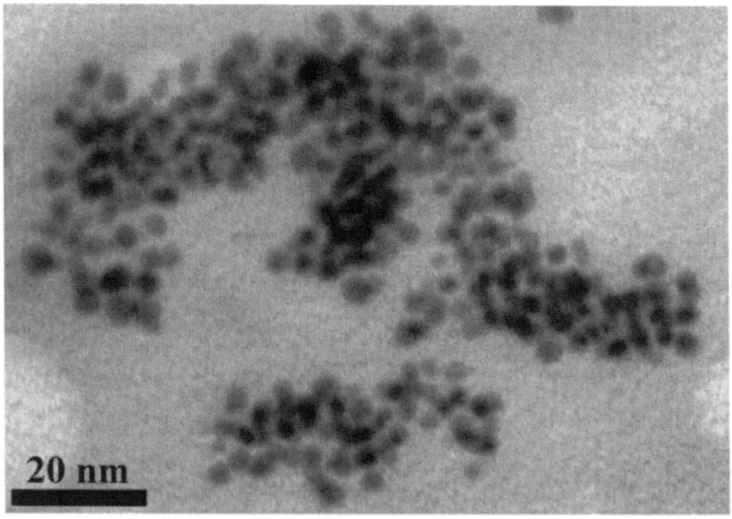

Figure 3. TEM observation of p-aminothiophenol-functionalized gold nanoparticles.

The FTIR spectrum of p-aminothiophenol was compared to that of p-aminothiophenol-functionalized gold nanoparticles (cf Figure 4). The main peaks of p-aminothiophenol are present in both FTIR transmission spectra (1202 cm^{-1}, 1489 cm^{-1}, 1589 cm^{-1}).

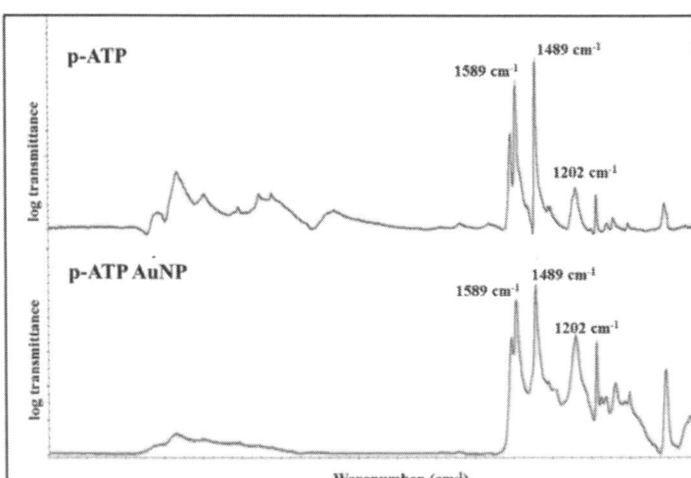

Figure 4. FTIR spectrum of p-aminothiophenol (ATP) and of p-aminothiophenol-functionalized gold nanoparticles (Au-ATP).

2.1.2. Electropolymerization of AFB1 MIP-MOF Film

The first step to prepare the electrodes was the self-assembly of PATP on the gold electrode surface by Au-S bonds and the subsequent deposition of an electropolymerized layer of polythioaniline/AuNPs/aflatoxin. The aflatoxin template was then removed from the matrix, providing cavities that can recognize and bind aflatoxin in the next step.

The electropolymerizaton was performed by CV in a solution of 10 mM $[Fe(CN)_6]^{3-/4-}$ in PBS pH 7.2 and the film formation was examined through the changes in the current per cycle. As shown in Figure 5, the current decreased as the number of cycles increased, suggesting gradual growth of a compact polymeric film which covered the surface of the gold electrode, hindering the charge transfer of the redox probe at the electrode. The electrochemical polymerization process consists of intermolecular reactions between the aniline moieties of PATP.

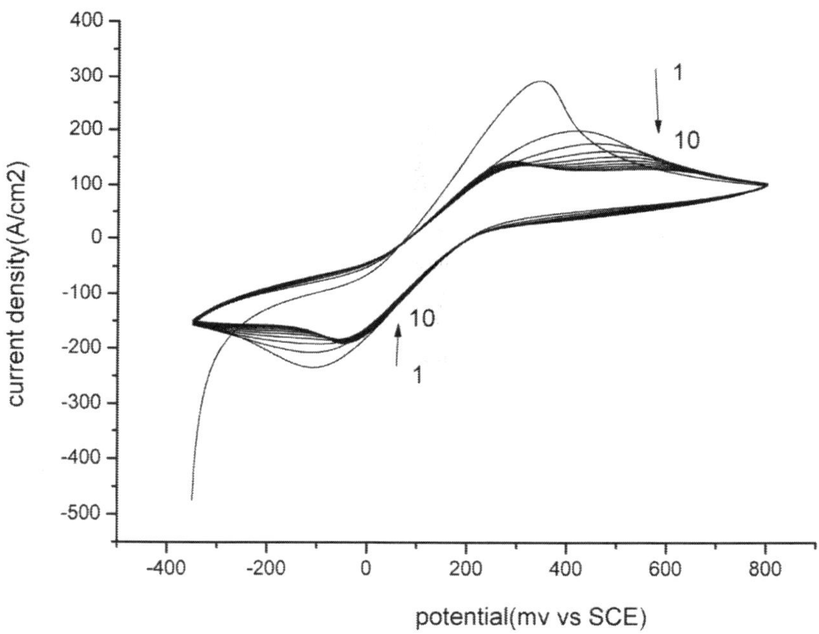

Figure 5. Electropolymerization of MIP by CV. Potential range from −0.35 V to 0.8 V vs. SCE. Scan rate 100 mV/s.

2.1.3. Structural and Electrochemical Characterization of AFB1 MIP-MOF Film

Figure 6 shows the phase of the prepared film, observed by AFM, which offers important information about the elasticity modulus of the MIP MOF film surface. This hybrid polymer film presents a very high phase variation: 24 degrees. This variation is due to the apparent AuNPs, their apparent diameter being around 30 nm, which is larger than that measured with TEM, due to the curvature radii of the silicon tip used.

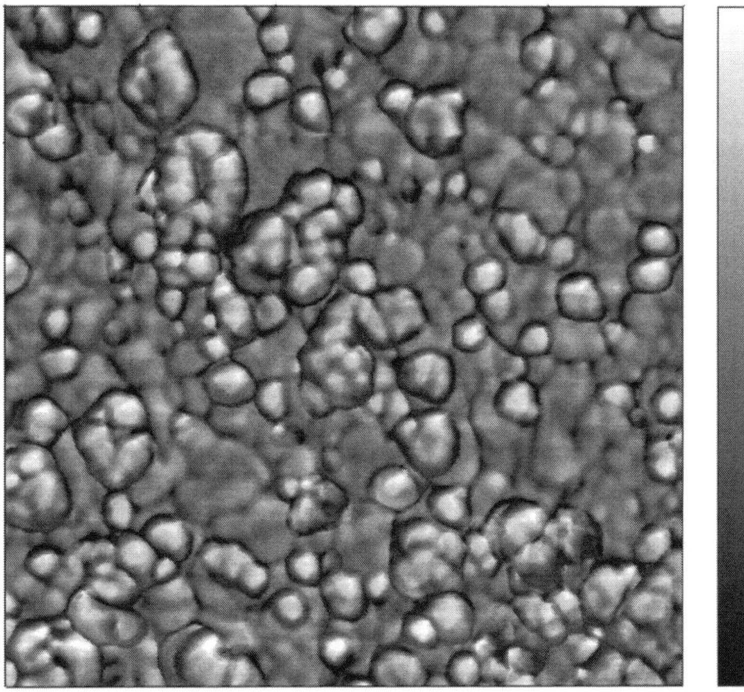

Figure 6. AFM phase image showing gold nanoparticles included in the MIP MOF film. Size of image: 1 µm × 1 µm; Vertical scale: phase from −10.3 degrees to +14.3 degrees.

After the electropolymerization, MIP and NIP films were characterized using linear sweep voltammetry (LSV), $[Fe(CN)_6]^{3-/4-}$ being used as a redox mediator. Their electrochemical behavior is quite different, as shown by the voltammograms presented in Figure 7. Charge transfer through NIP film is very low, compared to that through MIP. This shows that their morphology should be quite different and that the presence of a template molecule (AFB1) could help the charge transfer.

2.2. Recognition Ability of MIP MOF Films toward Aflatoxin B1

2.2.1. Electrochemical Detection of AFB1 Recognition by MIP-MOF Film

In order to check the uptake process for variable concentrations of aflatoxin B1, LSV measurements were performed in a solution of PBS pH 7.2 containing 10 mM $[Fe(CN)_6]^{3-/4-}$. After the extraction of the template, the prepared MIP was incubated with solution of different concentrations of aflatoxin B1. It was observed that an increase in the current peak and a shift of the peak potentials to a more negative potential with the increase of aflatoxin concentration from 3.2 fM to 3.2 μM (Figure 8) related to the binding of the analyte in the specific cavities. The higher concentration of aflatoxin molecules in MMOF film increases the electron transfer rate.

2.2.2. Mechanism of AFB1 Recognition by MIP-MOF Film

There are π-donor/π-acceptor interaction between aniline moieties and aflatoxin B1 molecules (due to dione functions of AFB1 molecule). The conductivity of the MIP-MOF film is partly based on polythioaniline, a conjugated polymer electron donor. This conductivity is increased when the concentration of AFB1 molecules in the polymer increases. A p-doping of the conjugated polymer is then induced, due to charge delocalization, creating a discrete donor level in the energy band diagram of polythioaniline [24]. This phenomenon can explain how the electron transfer rate of the ferro/ferricyanide redox probe is increased when the concentration of AFB1 increases. This phenomenon has been evidenced through fluorescence quenching [21] and through increase of electron transfer rate for a MIP-MOF film when concentration of nitroaromatic molecules increases [22].

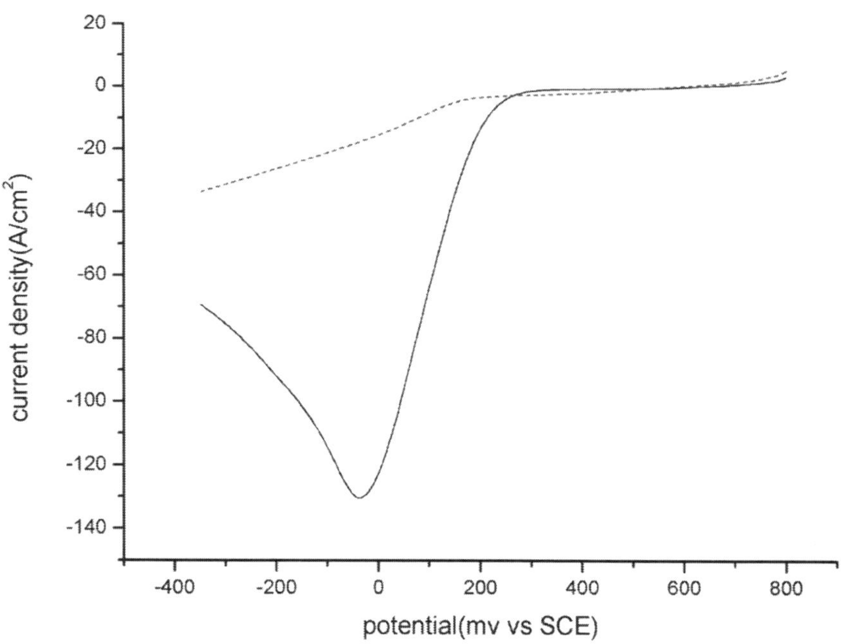

Figure 7. LSV behavior of the MIP (black line) and NIP (dotted red line) sensors after electropolymerization.

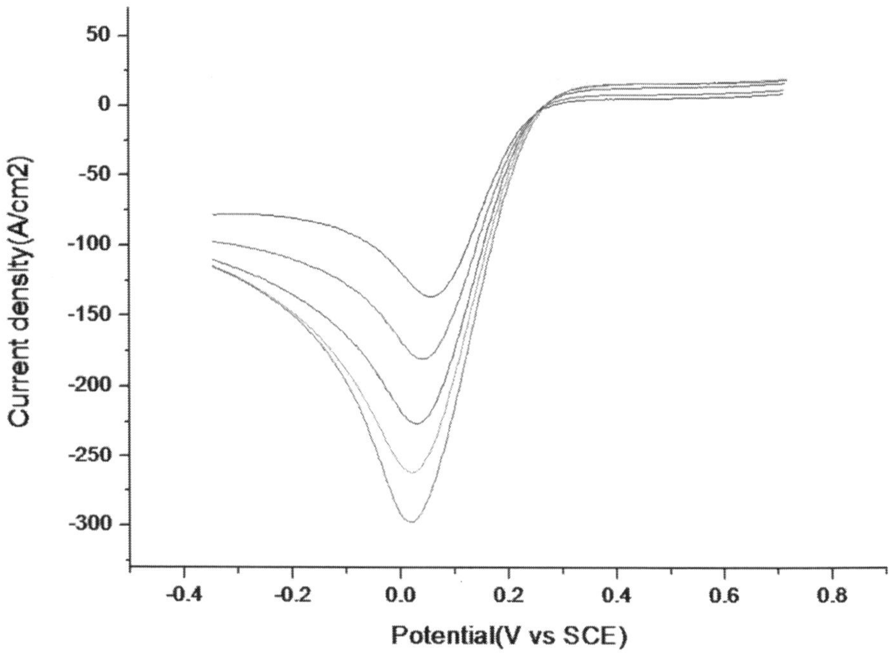

Figure 8. LSV after extraction of the template (black curve) and upon rebinding of various concentrations of aflatoxin: 3.2×10^{-15} M (red curve), 3.2×10^{-12} M (blue curve), 3.2×10^{-9} M (green curve), 3.2×10^{-6} M (purple curve). Potential range: 0.8 V to −0.35 V vs. SCE. Scan rate 50 mV/s.

2.3. Analytical Performance of the MIP MOF Sensor

From Figure 9 we can see the dependence of LSV current density on the concentration of aflatoxin. Data points were calculated as the average of three different electrodes. The peak current density ratio (signal-blank)/blank versus logarithm function of aflatoxin concentration exhibited a linear response over a range from 3.2 fM to 3.2 μM, with a limit of quantification of 3 fM (1 pg/L) and a limit of detection of 1 fM (0.3 pg/L). The limit of quantification (LOQ) was determined to be 10 σ_b and the limit of detection as 3 σ_b, σ_b being the standard deviation of the blank. The low signals obtained for NIP prove the influence of the template in obtaining binding sites for aflatoxin. The imprinting factor (ratio of sensitivity for MIP to sensitivity for NIP) is equal to 11.

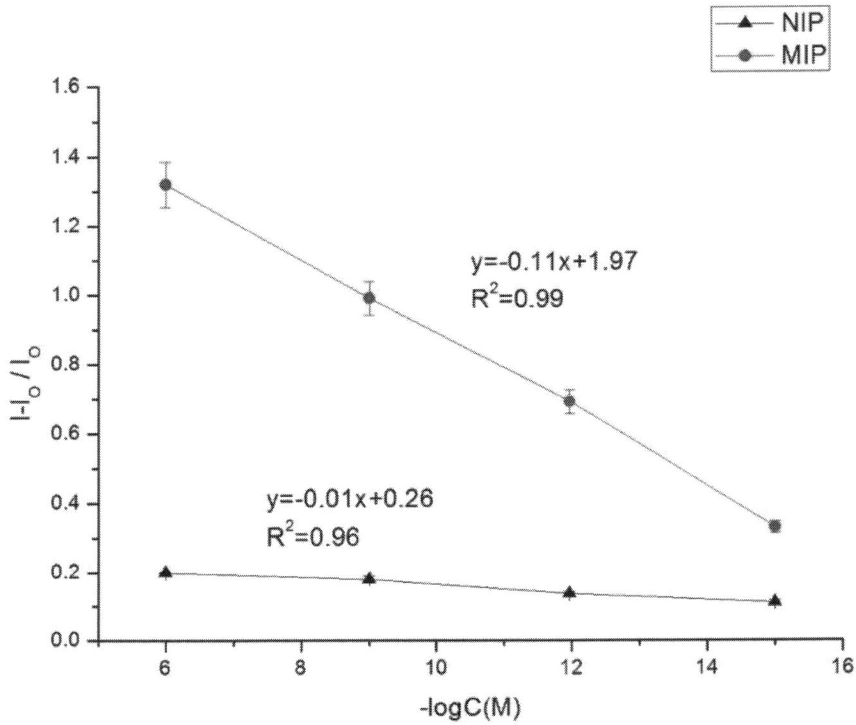

Figure 9. Calibration curves obtained for MIP (red curve) and NIP (brown curve) sensors by LSV for 3.2 fM to 32 nM aflatoxin. Under optimized parameters: extraction time 30 min, incubation time 20 min. I = signal after incubation with aflatoxin. I_O = signal after extraction.

As shown in Table 1, compared to previously reported methods (electrochemistry, HPLC/MS), the proposed sensor exhibits better analytical performance in terms of sensitivities.

The fabrication reproducibility was examined with three different modified electrodes, which were constructed independently using the same procedure. LSV studies were performed on each electrode aforementioned for the binding of 3.2 pM of aflatoxin. The precision of the described procedure, in terms of RSD, was 5.6%, which indicated good reproducibility of the fabrication method.

To check the reusability of the proposed sensor, several cycles of extraction and incubation with a solution of 3.2 pM aflatoxin were performed on the same electrode. The sensor maintained its response for three cycles with good reproducibility (RSD 4%). When used in real sample analysis (spiked rice samples), a thorough washing of the sensor surface in a PBS buffer, pH 7.3, is necessary for the complete elimination of the matrix effect.

Table 1. Comparison of different methods for the detection of Aflatoxin B1.

Method	LOD	Linear Range	Reference
Immunosensor on screen-printed carbon electrodes	90 pg/mL	0.1–10 ng/mL	[25]
Immunosensor on screen-printed microplate	30 pg/mL	0.05–2 ng/mL	[26]
Immunosensor on glassy carbon electrode	0.07 ng/mL	0.6–2.4 ng/mL	[27]
HPLC	0.25 ng/mL	0.5–10 ng/mL	[28]
HPLC-MSMS	0.08 ng/mL	0.3–10 ng/mL	[29]
LC-MSMS	2.4 pg/mL	0.05–2 ng/mL	[6]
LSV on molecularly imprinted MOF	0.3 fg/mL	1 fg/mL–1 μg/mL	Present method

Selectivity is a determining parameter in the analytical features of the MIP-MOF sensor. The responses for mycotoxins of similar structures, such as aflatoxin B2 (AFB2), aflatoxin G1 (AFG1), and ochratoxin A (OTA) have been examined. Figure 10 displays the response for different concentrations of interfering species; the peak current density ratio (signal-blank)/blank versus logarithm of different concentrations exhibit a linear response over a range from 3.2 fM to 3.2 μM. The cross-reactivity in percent of the MIP-MOF sensor for AFB2, AFG1, and OTA is reported in Table 2, as one hundred times the ratio of concentration of AFB1 giving a response of MIP-MOF sensor to the concentration of other mycotoxin giving the same response. The results show that the proposed biosensor has high selectivity for aflatoxin B1.

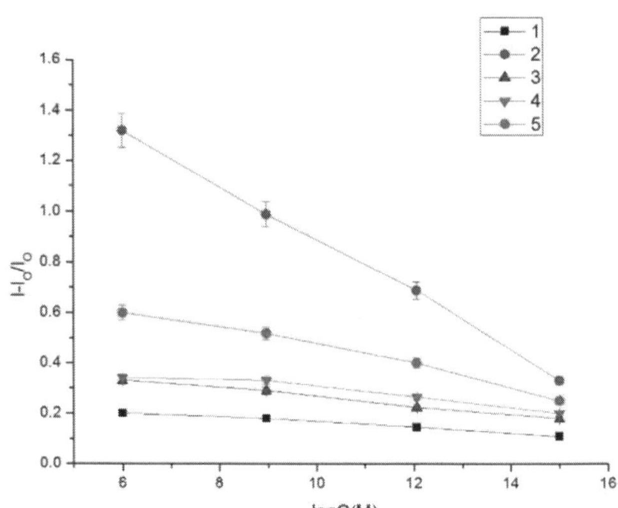

Figure 10. Calibration curves upon analysis of various concentrations of aflatoxins and ochratoxin A. Curve 1: NIP sensor for AFB1; Curve 2: MIP sensor for AFB1; Curve 3: MIP sensor for AFB2; Curve 4: MIP sensor for AFG1; Curve 5: MIP sensor for OTA.

Table 2. Cross-reactivity of aflatoxins and ochratoxin versus AFB1.

Type of Mycotoxin	Cross-reactivity (%)
Aflatoxin B2	10^{-7}
Aflatoxin G1	10^{-7}
Ochratoxin A	10^{-5}

3. EXPERIMENTAL SECTION

3.1. Chemicals and Instrumentation

p-Aminothiophenol (PATP), tetrachloroauric acid trihydrate ($HAuCl_4 \cdot 3H_2O$), potassium ferrocyanide ($K_4[Fe(CN)_6]$), sodium borohydride ($NaBH_4$), potassium ferricyanide ($K_3[Fe(CN)_6]$), methanol, and phosphate buffered saline (PBS) tablets were purchased from Sigma-Aldrich (Saint-Quention Fallavier, France). All solutions were prepared using Milli-Q water with a resistivity of 18.2 MΩ·cm at 25 °C. Aflatoxin B1 (1 mg), Aflatoxin B2 (1 mg), Aflatoxin G1 (1 mg), and Ochratoxin A (1 mg) were purchased from Sigma-Aldrich.

UV-VIS absorption spectra of the AuNPs nanoparticles were recorded at room temperature with a Lambda 900 UV/Vis/NIR spectrophotometer (Perkin Elmer, Waltham, MA, USA) between 300 and 800 nm.

Transmission electron microscopy (TEM) pictures were obtained using a CM120 TEM (Philips, Amsterdam, Netherlands) with an accelerating voltage of 120 kV (Centre Technologique des Microstructures-Lyon 1, Villeurbanne, France). AuNPs were examined after deposition of 5 mL of diluted solutions on a formvar-coated copper grid (Electron Microscopy Sciences) and evaporation to dryness.

Infrared spectra were recorded at room temperature using a Continuμm microscope coupled with Nexus infrared spectroscopy in specular reflectance equipped with an MCT detector (Nicolet Instruments Corporation, Madison, WI, USA). Recordings were obtained with a resolution of 4 cm^{-1}, a spectral width between 690 and 4000 cm^{-1} and signal processing through Happgenzel apodization. (256 scans).

Electrochemical measurement were performed with a VOLTALAB-80 PGZ/301 potentiostat-galvanostat (Hach, Lange, France), controlled by Voltamaster 4 software. Experiments employed in a three-electrode electrochemical cell, with a gold working electrode (0.07 cm^2), a platinum plate (0.17 cm^2) as auxiliary electrode and a saturated calomel electrode(SCE) as reference.

AFM was used to investigate the surface topography of the electropolymerized films and their rheological response (phase). The AFM measurements were carried out using an Agilent 5500 AFM (Agilent Technologies, Palo Alto, CA, USA). Silicon tips with a nominal spring constant of 20 Nm^{-1} were used in tapping mode at a frequency of ~300 kHz.

3.2. Preparation of PATP Functionalized AuNPs

The AuNPs were synthesized according to the literature method [30]. A typical preparation of gold nanoparticles is as follows: 31.6 mg (8×10^{-5} mol) of tetrachloroauric (III) acid trihydrate was dissolved in 30 mL of methanol in a round 100 mL flask equipped with a condenser. A solution of p-aminothiophenol (1.6×10^{-4} mol) in 12 mL of a methanol/water (v/v) mixture was added dropwise under stirring to the gold salt solution, which changed color from yellow to dark brown. After 10 min, 30.4 mg of $NaBH_4$ (8×10^{-4} mol) dissolved in 2.2 mL of water was added dropwise to the mixture under vigorous stirring. After 10 min, stop stirring and keep the solution in darkness for 1 h. The suspension was then filtered through a polymer membrane and washed successively with water and ether. The resulting black powder was dried and stored either as a solid or dispersed in 0.1 N HCl solution. The UV-vis spectra of gold nanoparticle colloids show a small surface plasmon band centered around 535 nm. The solid PATP-fuctionalized AuNPs were characterized using TEM and FTIR.

3.3. Preparation of MIP and Non-Imprinted (NIP) MOF Modified Electrodes

Both MIP and NIP films were prepared on the surface of gold electrodes using cyclic voltammetry (CV), by electropolymerization of PATP self-assembled on gold electrodes and PATP functionalized gold nanoparticles. Gold substrates were provided by the French RENATECH network (LAAS, CNRS Toulouse, France). They were fabricated using standard silicon technologies. (100)-oriented, P-type (3–5 Ω·cm) silicon wafers were thermally oxidized to grow an 800 nm-thick field oxide. Then, a 30 nm-thick titanium layer followed by a 300 nm thick gold top layer were deposited by evaporation under vacuum. The gold electrodes were 1×1 cm^2 square plates. Prior to SAM formation, the gold electrodes were cleaned by rinsing with ethanol and water followed by exposure to UV/ozone. An intermediate monolayer of PATP was firstly formed by immersing the gold electrodes in a 50 mM PATP ethanol solution for 12 h at 4 °C. After rinsing with ethanol and water to remove the physically adsorbed PATP, the electropolymerizaton of MIP was conducted in a solution containing 0.1 mg/mL PATP functionalized AuNPs and 0.1 mg/mL aflatoxin B1 in the supporting electrolyte (10 mM $Fe(CN)_6^{3-/4-}$ in PBS pH 7.2). The potential was cycled between −0.3 V and +0.80 V vs. SCE, at a scan rate of 100 mv/s, for 10 cycles. A control non-imprinted film (NIP) was also prepared in every case, with the same method, but without any additional aflatoxin B1.

The modified electrodes were then washed with a 1:4 (v/v) mixture of methanol/PBS solution for 30 min to remove the imprinted molecules and adsorbed molecules on the surface of the film, triggering the formation of recognition cavities in the MIP film.

In order to promote the recognition and binding of AFB1 molecules, the MIP electrode was left to incubate with aflatoxin B1 solutions for 20 min. For the characterization of the MIP sensor at different steps of the preparation, electrochemical measurements were carried out in 10 mM $[Fe(CN)_6]^{3-/4-}$ in PBS pH 7.2 at room temperature. Linear sweep voltammetry was employed from 0.80 V to −0.35 V at a scan rate of 50 mV/S.

4. CONCLUSIONS

The development of an electrochemical sensor for sensitive detection of aflatoxin B1 based on an MIP-based polymer/AuNPs hybrid membrane is reported for the first time herein. The MIP was prepared on the surface of PATP-functionalized gold electrodes by electropolymerization of PATP-functionalized AuNPs in the presence of AFB1 as the template molecule. The developed sensor showed a wide linear range between 3.2 fM and 3.2 μM with an LOQ of 3 fM, lower than that reported for other AFB1 sensors. With good sensitivity and reproducibility, the MIP hybrid sensor was shown to have the potential to be an effective method for the selective electrochemical detection of AFB1 in food samples. The easy fabrication and robustness of molecularly-imprinted MOF films make them ideal candidates for the development of sensing devices.

ACKNOWLEDGMENTS

The authors thank for financial support EC through the Marie Curie NANODEV Grant No318524.

REFERENCES

1. Šegvić Klarić, M.; Rašić, D.; Peraica, M. Deleterious Effects of Mycotoxin Combinations Involving Ochratoxin A. Toxins **2013**, 5, 1965–1987.

2. Joint FAO/WHO Expert Committee on Food Additives (JECFA). Safety Evaluation of Certain Food Additives and Contaminants, WHO Food Additives Series 59; WHO: Geneva, Switzerland, 2008; pp. 305–356.

3. Balzer, I.; Bogdanic, C.; Pepeljnjak, S. Rapid thin layer chromatographic method for determining aflatoxin B1, ochratoxin A, and zearalenone in corn. J. Assoc. Off. Anal. Chem. **1978**, 61, 584–585.

4. Hu, Y.Y.; Zheng, P.; Zhang, Z.X.; He, Y.Z. Determination of aflatoxins in high-pigment contents samples by matrix solid-phase dispersion and high-performance liquid chromatography. Asia-Australas. J. Anim. Sci. **2015**, 28, 691–696.

5. Sun, D.D.; Gu, X.; Li, J.G.; Yao, T.; Dong, Y.C. Quality Evaluation of Five Commercial Enzyme Linked Immunosorbent Assay Kits for Detecting Aflatoxin B1 in Feedstuffs. Biosens. Bioelectron. **2007**, 22, 1434–1440.

6. Nonaka, Y.; Saito, K.; Hanioka, N.; Narimatsu, S.; Kataoka, H. Determination of aflatoxins in food samples by automated on-line in-tube solid-phase microextraction coupled with liquid chromatography-mass spectrometry. J. Chromatogr. A **2009**, 1216, 4416–4422.

7. Diltemiz, S.E.; Hur, D.; Ersoz, A.; Denizli, A.; Say, R. Designing of MIP based QCM sensor having thymine recognition sites based on biomimicking DNA approach. Biosens. Bioelectron. **2009**, 25, 599–603.

8. Alizadeh, T.; Akhoundian, M. A novel potentiometric sensor for promethazine based on a molecularly imprinted polymer (MIP): The role of MIP structure on the sensor performance. Electrochim. Acta **2010**, 55, 3477–3485.

9. Li, J.; Li, S.; Wei, X.; Tao, H.; Pan, H. Molecularly Imprinted Electrochemical Luminescence Sensor Based On Signal Amplification for Selective Determination of Trace Gibberellin A3. Anal. Chem. **2012**, 84, 9951–9955.

10. Suriyanarayanan, S.; Cywinski, P.J.; Moro, A.J.; Mohr, G.J.; Kutner, W. Chemosensors Based on Molecularly Imprinted Polymers. In Molecular Imprinting; Haupt, K., Ed.; Springer: Berlin/Heidelberg, Germany, 2012; Volume 325, pp. 165–265.

11. Schirhagl, R. Bioapplications for Molecularly Imprinted Polymers. Anal. Chem. **2013**, 86, 250–261.

12. Yano, K.; Karube, I. Molecularly imprinted polymers for biosensor applications. Trac. Trends Anal. Chem. **1999**, 18, 199–204.

13. Tokonami, S.; Shiigi, H.; Nagaoka, T. Review: micro- and nanosized molecularly imprinted polymers for high-throughput analytical applications. Anal. Chim. Acta **2009**, 641, 7–13.

14. Moreno-Bondi, M.C.; Benito-Pena, M.E.; Urraca, J.L.; Orellana, G. Immuno-Like Assays and Biomimetic Microchips. In Molecular Imprinting; Haupt, K., Ed.; Springer: Berlin/Heidelberg, Germany, 2012; Volume 325, pp. 111–164.

15. Apodaca, D.C.; Pernites, R.B.; Ponnapati, R.R.; Del Mundo, F.R.; Advincula, R.C. Electropolymerized molecularly imprinted polymer films of a bis-terthiophene dendron: folic acid quartz crystal microbalance sensing. ACS Appl. Mater. Interfaces **2011**, 3, 191–203.

16. Yaghi, O.M.; Li, H.L. Hydrothermal synthesis of a metal organic framework containing large rectangular channels. J. Am. Chem. Soc. **1995**, 117, 10401–10402.

17. Yaghi, O.M.; Li, G.M.; Li, H.L. Selective binding and removal of guests in a microporous metal-organic framework. Nature **1995**, 378, 703–706.

18. Pan, L.; Parker, B.; Huang, X.; Olson, D.H.; Lee, J.; Li, J. Zn(tbip) (H_2tbip = 5-tert-Butyl Isophthalic Acid): A Highly Stable Guest-Free Microporous Metal Organic Framework with Unique Gas Separation Capability. J. Am. Chem. Soc. **2006**, 128, 4180–4181.

19. Morozan, A.; Jaouen, F. Metal organic frameworks for electrochemical applications. Energy Environ. Sci. **2012**, 5, 9269–9290.

20. Ma, T.Y.; Li, H.; Deng, Q.F.; Liu, L.; Ren, T.Z.; Yuan, Z.Y. Ordered Mesoporous Metal-Organic Frameworks Consisting of Metal Disulfonates. Chem. Mater. **2012**, 24, 2253–2255.

21. Pramanik, S.; Zheng, C.; Zhang, X.; Emge, T.J.; Li, J. New Microporous Metal-Organic Framework Demonstrating Unique Selectivity for Detection of High Explosives and Aromatic Compounds. J. Am. Chem. Soc. **2011**, 133, 4153–4155.

22. Guo, Z.; Florea, A.; Cristea, C.; Bessueille, F.; Vocanson, F.; Goutaland, F.; Zhang, A.; Săndulescu, R.; Lagarde, F.; Jaffrezic-Renault, N. 1,3,5-Trinitrotoluene detection by a molecularly imprinted polymer sensor based on electropolymerization of a microporous-metal-organic framework. Sens. Actuators B **2015**, 207, 960–966.

23. Florea, A.; Guo, Z.; Cristea, C.; Bessueille, F.; Vocanson, F.; Goutaland, F.; Dzyadevych, S.; Săndulescu, R.; Jaffrezic-Renault, N. Anticancer drug detection using a highly sensitive molecularly imprinted electrochemical sensor based on an electropolymerized microporous metal organic framework. Talanta **2015**, 138, 71–76.

24. Mane, A.T.; Navale, S.T.; Patil, V.B. Room temperature NO_2 gas sensing properties of DBSA doped PPy-WO_3 hybrid nanocomposite sensor. Org. Electron. **2015**, 19, 15–25.

25. Ammida, N.; Micheli, L.; Piermarini, S.; Moscone, D.; Palleschi, G. Detection of aflatoxin B1 in barley: Comparative study of immunosensor and HPLC. Anal. Lett. **2006**, 39, 1559–1572.

26. Piermarini, S.; Micheli, L.; Ammida, N.H.S.; Palleschi, G.; Moscone, D. Electrochemical immunosensor array using a 96-well screen-printed microplate for aflatoxin B1 detection. Biosens. Bioelectron. **2007**, 22, 1434–1440.

27. Owino, J.H.; Arotiba, O.A.; Hendricks, N.; Songa, E.A.; Jahed, N.; Waryo, T.T.; Ngece, R.F.; Baker, P.G.; Iwuoha, E.I. Electrochemical Immunosensor Based on Polythionine/Gold Nanoparticles for the Determination of Aflatoxin B1. Sensors **2008**, 8, 8262–8274.

28. Hu, Y.Y.; Zheng, P.; Zhang, Z.X.; He, Y.Z. Determination of Aflatoxins in High-Pigment Content Samples by Matrix Solid-Phase Dispersion and High-Performance Liquid Chromatography. J. Agric. Food Chem. **2006**, 54, 4126–4130.

29. Han, Z.; Zheng, Y.; Luan, L.; Cai, Z.; Ren, Y.; Wu, Y. An ultra-high-performance liquid chromatography tandem mass spectrometry method for simultaneous determination of aflatoxins B1, B2, G1, G2, M1 and M2 in traditional Chinese medicines. Anal. Chim. Acta **2010**, 664, 165–171.

30. Debouttière, P.J.; Roux, S.; Vocanson, F.; Billotey, C.; Beuf, O.; Favre-Réguillon, A.; Lin, Y.; Pellet-Rostaing, S.; Lamartine, R.; Perriat, P.; et al. Design of Gold Nanoparticles for Magnetic Resonance Imaging. Adv. Funct. Mater. **2006**, 16, 2330–2339.

Chapter 4

Electrodeposition and Capacitive Behavior of Films for Electrodes of Electrochemical Supercapacitors

C. Shil. Zhitomirsky

Department of Materials Science and Engineering, McMaster University, 1280 Main Street West, Hamilton, ON L8S 4L7, Canada

ABSTRACT

Polypyrrole films were deposited by anodic electropolymerization on stainless steel substrates from aqueous pyrrole solutions containing sodium salicylate and tiron additives. The deposition yield was studied under galvanostatic conditions. The amount of the deposited material was varied by the variation of deposition time at a constant current density. SEM studies showed the formation of porous films with thicknesses in the range of 0–3 μm. Cyclic voltammetry data for the films tested in 0.5 M Na_2SO_4 solutions showed capacitive behavior and high specific capacitance (SC) in a voltage window of 0.9 V. The films prepared from pyrrole solutions containing tiron showed better capacitive behavior compared to the films prepared from the solutions containing sodium salicylate. A highest SC of 254 F g^{-1} was observed for the sample with a specific mass of 89 μg cm^{-2} at a scan rate of 2 mV s^{-1}. The SC decreased with an increasing film thickness and scan rate. The results indicated that the polypyrrole films deposited on the stainless steel substrates by anodic electropolymerization can be used as electrodes for electrochemical supercapacitors (ES).

Keywords: Polypyrrole Film Sodium salicylate Tiron Supercapacitor Capacitance Electropolymerization

INTRODUCTION

The development of hybrid and electric vehicles requires the use of efficient ES [1], which provide load-levelling during starting, acceleration and braking. Polypyrrole is an attractive material for the fabrication of electrodes for ES due to its high SC and good electrical conductivity. Numerous investigations have been conducted with a goal of utilizing the high SC of polypyrrole in ES. Impressive progress has already been made in the fabrication and testing of polypyrrole films prepared by various methods, such as emulsion polymerization [2], layer-by-layer assembly [3], chemical [4] and electrochemical polymerization [5]. The mechanism of charge storage in the polypyrrole electrodes has been investigated using quartz crystal microbalance method [5].

Electrochemical polymerization is an attractive method for the fabrication of polypyrrole films. In this approach, anodic polymerization of polypyrrole can be achieved from aqueous monomer solutions containing anionic additives. Various substrates were used for the anodic electropolymerization of polypyrrole films, including tantalum [6], lead [7], titanium [8], platinum [9], indium tin oxide [10, 11], porous carbon fibre paper [12], activated carbon [13], graphite [14] and carbon foam [15]. The films exhibited excellent capacitive behavior with a SC as high as 400 F g^{-1} and good cycling stability during 10,000 cycles [14]. However, the deposition of polypyrrole on low-cost stainless steel substrates presents difficulties related to anodic oxidation and dissolution of the substrates. The chemical and electrochemical passivation of the substrates for electropolymerization of polypyrrole usually results in the formation of insulating films with low capacitance. The formation of such films results in a reduced total capacitance of the electrodes. The polypyrrole films [16] deposited on the stainless steel substrate using p-toluene sulphonic acid as an additive showed a SC of ~100 F g^{-1}.

Recently, it was shown that salicylic acid and tiron are promising additives for the electrochemical polymerization of polypyrrole [17–19]. Strongly adherent films of polypyrrole were obtained on zinc-electroplated stainless steel using sodium salicylate as an additive [17]. It was shown that sodium salicylate complexes the surface metal ions and prevents the dissolution of metallic substrates. However, such surface complexation does not prevent electropolymerization of the polypyrrole. Electropolymerization of polypyrrole on aluminum alloy was achieved using tiron [19]. It was shown that the tiron acted as a charge transfer mediator, which reduced the deposition potential by nearly 500 mV compared to the deposition performed in the absence of mediator. The method enabled the formation of adherent films with good electrical conductivity.

The results presented below indicated that sodium salicylate and tiron are efficient processing additives for the fabrication of polypyrrole films on stainless steel substrates by anodic electropolymerization. We presented experimental data on the capacitive behavior and microstructure of the films prepared by this method.

EXPERIMENTAL PROCEDURES

Sodium salicylate, tiron and pyrrole were purchased from Aldrich. Electropolymerization was performed from an aqueous 0.25 M pyrrole solution containing 0.5 M sodium salicylate, or a 0.05 M pyrrole solution containing 0.005 M tiron. The deposition was performed galvanostatically at a current density of 1 mA cm^{-2}. The deposition cell contained a stainless steel foil (50 mm × 30 mm × 0.1 mm) or wire (diameter 0.1 mm) as working electrode and a platinum foil as counter electrode. Deposition time was varied in the range of 1–10 min. Deposition was also performed on platinized silicon wafers containing 1,000 Å Pt layer.

Electron microscopy investigations were performed using a JEOL JSM-7000F scanning electron microscope. The capacitive behavior of the electrodes was studied using a potentiostat (PARSTAT 2273, Princeton Applied Research) controlled by a computer using the PowerSuite electrochemical software. Electrochemical studies were performed using a standard three-electrode cell containing a 0.5 M Na_2SO_4 aqueous solution, degassed with purified nitrogen gas. The surface area of the working electrode was 1 cm^2. The counter electrode was a platinum gauze, and the reference electrode was a standard calomel electrode (SCE). Cyclic voltammetry (CV) studies were performed within a potential range of −0.5 to +0.4 V versus SCE at scan rates of 2–100 mV s^{-1}. The SC was calculated according to the following equation:

$$C = Q/m\Delta V \tag{1}$$

where Q is the charge obtained using half of the integrated area of the CV curve, m is the mass of the active material, and ΔV is the width of the potential window.

RESULTS AND DISCUSSION

Figure 1 shows the structures of sodium salicylate and tiron used in this study. Both materials were negatively charged in the aqueous solutions due to the dissociation of –COONa and $–SO_3Na$ groups. The materials served as anionic dopants, which were incorporated into the polymer to ensure the electrical neutrality of the growing film during the anodic polymerization of polypyrrole.

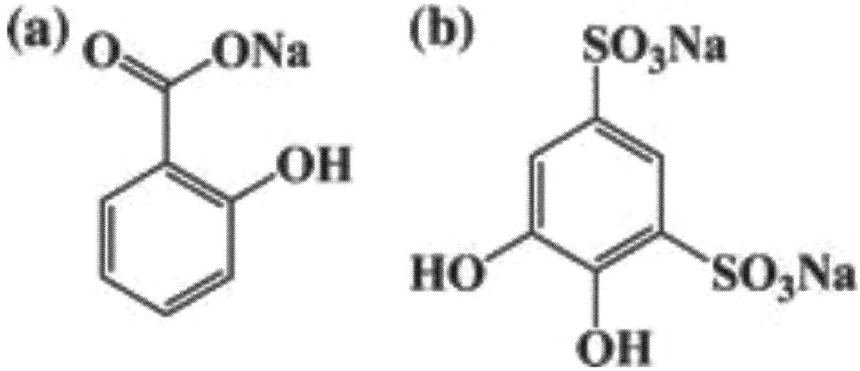

Figure 1. The chemical structure of **a** Sodium salicylate and **b** Tiron

Anodic polymerization from aqueous pyrrole solutions containing sodium salicylate or tiron resulted in the formation of adherent polypyrrole films on stainless steel substrates. The film mass increased with increasing deposition time (Fig. 2a, 2b), indicating the formation of films of different thicknesses. Nearly linear dependences were obtained. Therefore, the amount of the deposited materials can be controlled by the variation of deposition time at a constant current density. This approach can be used for film formation on other conductive substrates, such as Pt. SEM investigations (Fig. 3) of the cross-sections of the films on platinized silicon wafers showed that the film thickness was varied in the range of 0–3 μm. The films exhibited surface roughness, which can be attributed to the formation of polypyrrole particles. The films prepared from the pyrrole solutions containing tiron exhibited lower surface roughness compared to those prepared from the solutions containing sodium salicylate. The SEM images showed the formation of porous films. The film porosity is beneficial for application in ES, as it improves the electrolyte access to the electrochemically active electrode material [1].

Figure 4 shows surfaces of the films deposited on the stainless steel foils. Low-magnification images Fig. 4a, 4c showed that the films were crack free. The SEM image at a higher magnification, Fig. 4b, for the film prepared from the pyrrole solution containing sodium salicylate showed polypyrrole particles with a particle size of 0.1–0.5 μm. The film prepared from the solution containing tiron showed a much finer particle size (Fig. 4d). It is known from the literature [20] that the morphology of polypyrrole films depends on the nature of anionic additives. It is suggested that the lower roughness of the films prepared from solutions containing tiron can be attributed to the lower size of the polypyrrole particles.

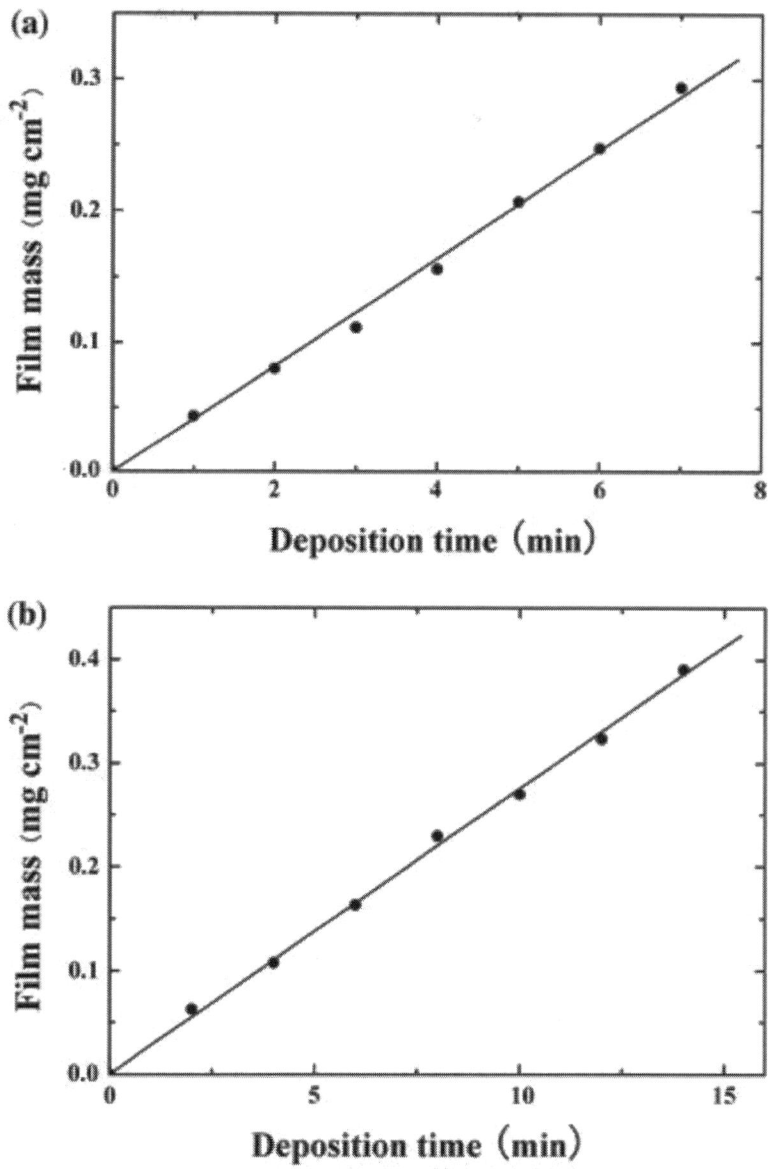

Figure 2. The film mass versus deposition time for the films prepared from the pyrrole solutions containing **a** Sodium salicylate and **b** Tiron on stainless steel foils.

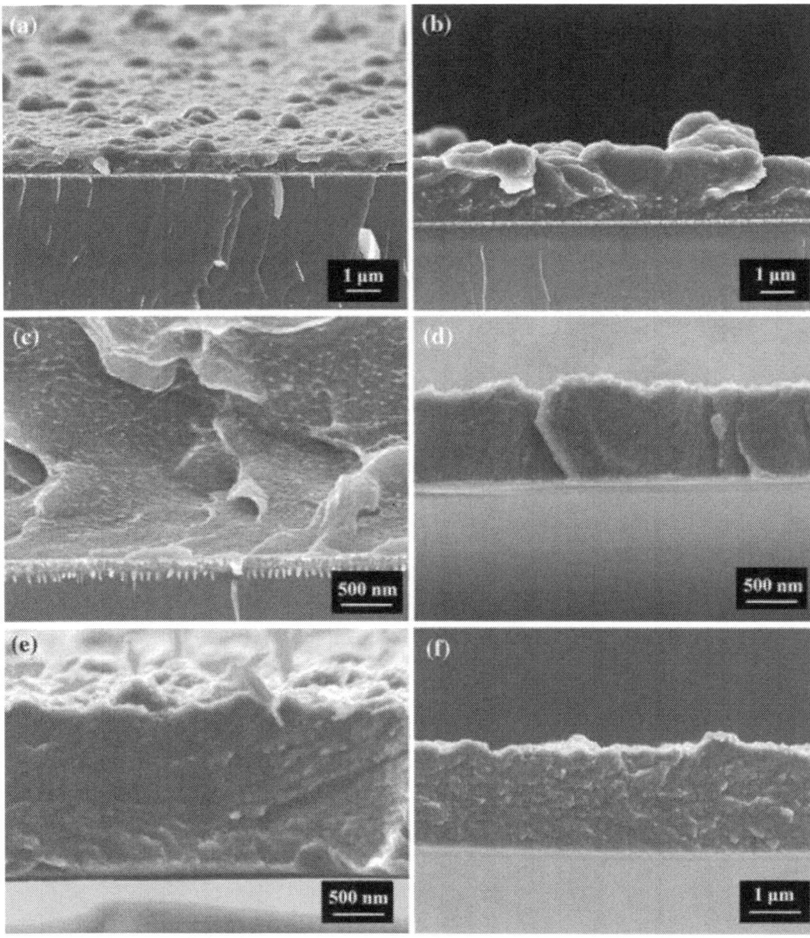

Figure 3. SEM images of cross-sections for the films of different thicknesses prepared from the pyrrole solutions containing **a**, **b**, **c** Sodium salicylate and **c**, **d**, **f** Tiron on platinized silicon wafers

Anodic electropolymerization has been utilized for the film formation on substrates of complex shapes. As an example, Fig. 5a shows a polypyrrole coated stainless steel wire. The SEM image of the cross-section indicated a significant surface roughness. The SEM image of the wire surface at a higher magnification showed that the film consisted of relatively large particles with a particle size of about 1 μm. Therefore, the surface roughness of the film can be attributed to the relatively large particle size, which was comparable with the film thickness. The films prepared from the pyrrole solutions containing tiron showed a better uniformity and lower surface roughness, which can be attributed to a lower particle size. Figure 6a, 6b shows typical images of polypyrrole films deposited on stainless steel wires from solutions containing tiron. The thickness of the films on the stainless steel wires was in the range of 0–3 μm.

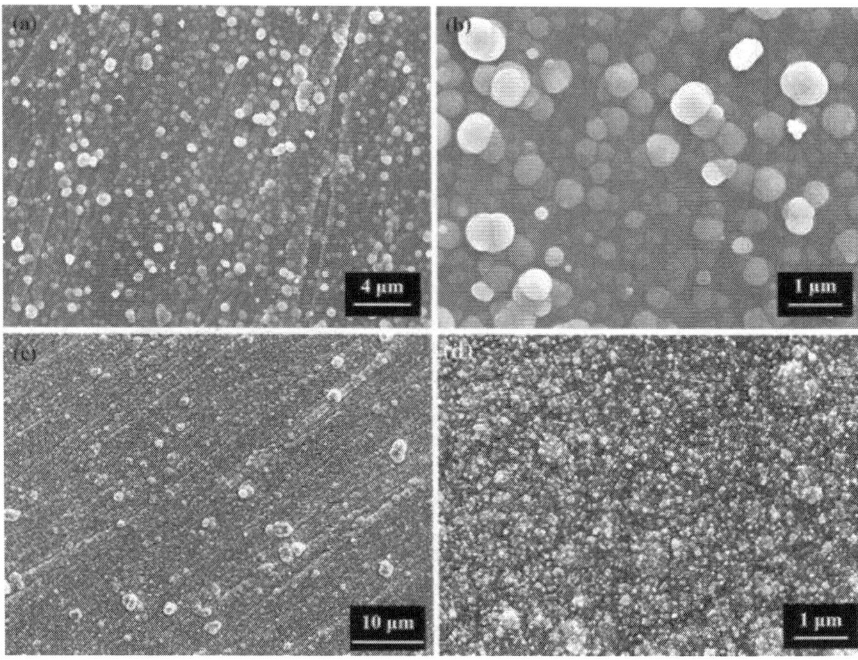

Figure 4. SEM images of surfaces for the films deposited from the pyrrole solutions containing **a**, **b** Sodium salicylate and **c**, **d** Tiron on stainless steel foils at different magnifications

The electrochemical properties of the films were studied in 0.5 M Na_2SO_4 solutions using CV. The films showed capacitive behavior in the voltage window of 0.9 V. However, the CVs for the films prepared from the solutions containing sodium salicylate deviated significantly from the ideal box shape (Fig. 7). The films prepared from the solutions containing tiron showed better capacitive behavior and exhibited box shape CVs (Fig. 8).

Figure 9 shows SC versus scan rate for the films of different thickness prepared from the pyrrole solutions containing sodium salicylate. The films showed SC in the range of 100–200 F g^{-1} at a scan rate of 2 mV s^{-1}. The SC decreased with increasing film thickness and increasing scan rate in the range of 2–100 mVs^{-1}. Such decrease was attributed to the electrolyte diffusion in the pores of polypyrrole films.

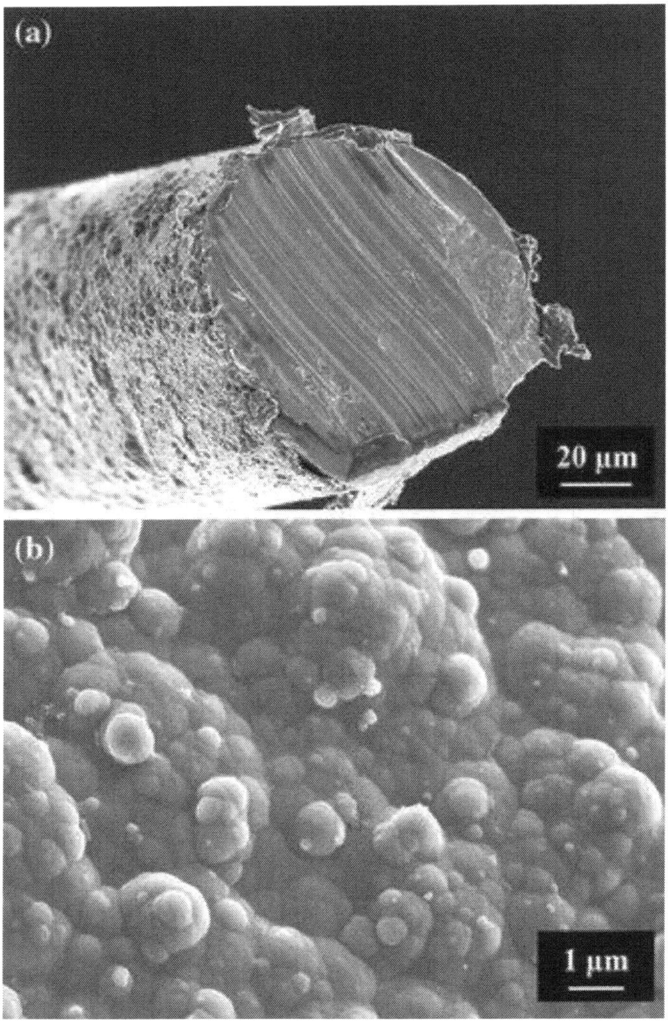

Figure 5. SEM images for the film deposited from the pyrrole solution containing sodium salicylate on a stainless steel wire

Figure 10 shows SC for films of different thickness prepared from the pyrrole solutions containing tiron. The highest SC of 254 F g^{-1} was observed for the 89 μm cm^{-2} sample at a scan rate of 2 mV s^{-1}. The SC decreased with increasing film thickness and increasing scan rate in the range of 2–100 mV s^{-1}. The films prepared from the solutions containing tiron showed higher SC compared to the films of the same mass prepared from solutions containing sodium salicylate. It is suggested that lower particle size of the polymer particles prepared from the solution containing tiron resulted in larger surface area of the particles and in better access of the electrolyte to the active material. As a result, the films prepared from the solutions containing tiron showed higher specific capacitance compared to the films prepared from the solutions containing sodium salicylate. However, the difference in the electrochemical behavior can also result from the different nature of the anionic groups of the additives.

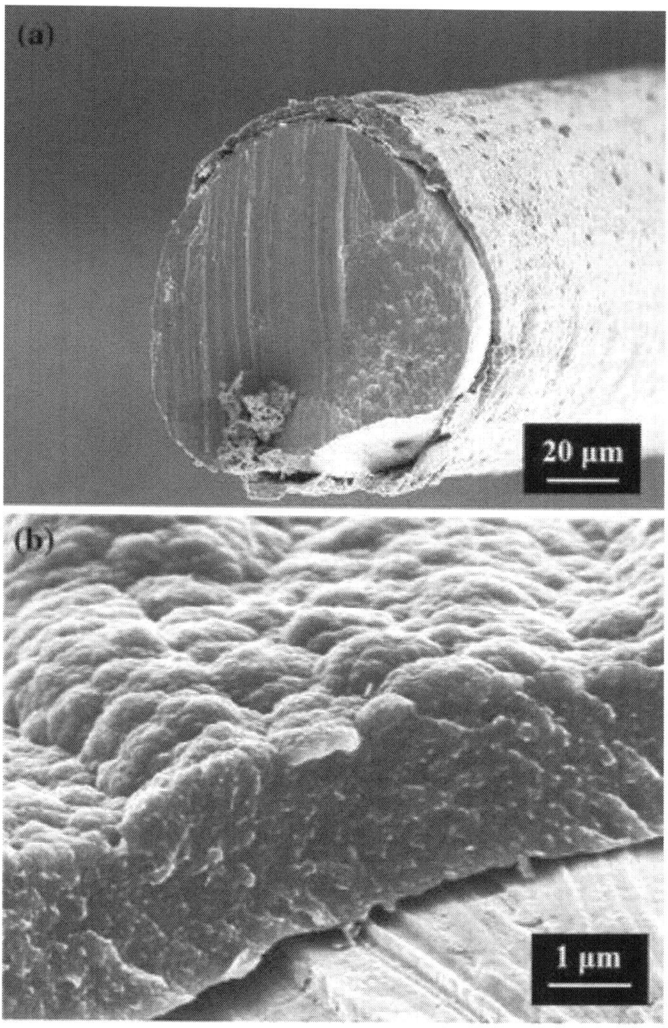

Figure 6 SEM images for the film deposited from the pyrrole solution containing tiron on a stainless steel wire

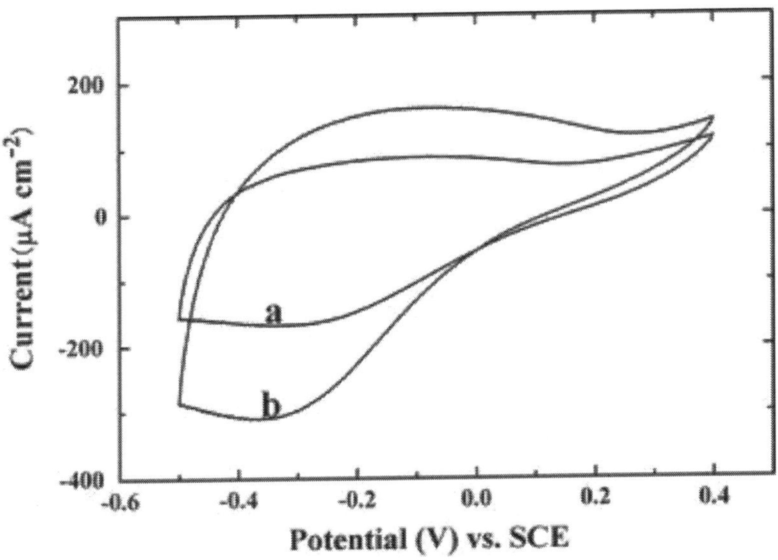

Figure 7. Cyclic voltammograms for the film deposited from the pyrrole solution containing sodium salicylate on a stainless steel foil at scan rates of (*a*) 5 and (*b*) 10 mV s^{-1}.

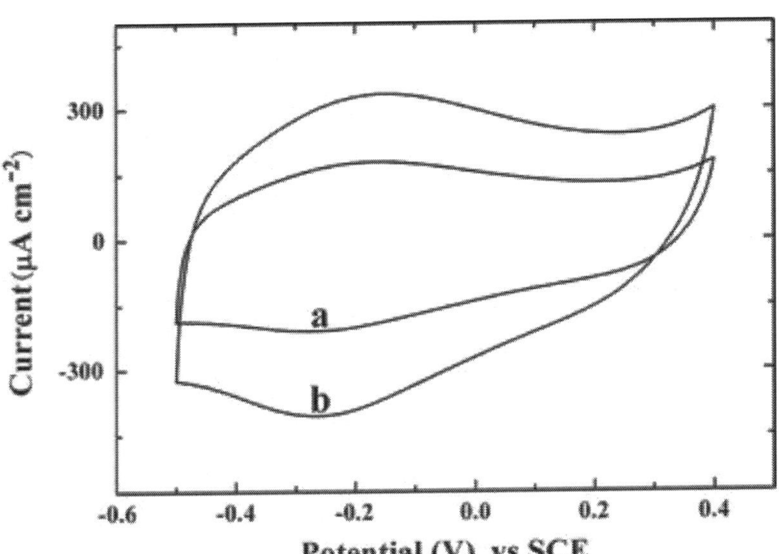

Figure 8. Cyclic voltammograms for the film deposited from the solution containing tiron on a stainless steel foil at scan rates of (*a*) 5 and (*b*) 10 mV s^{-1}.

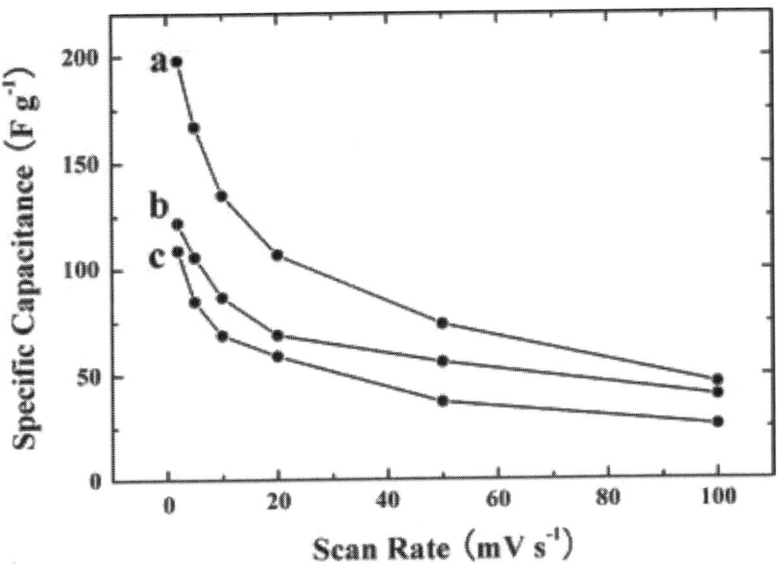

Figure 9. SC versus scan rate for the films deposited from the solution containing sodium salicylate on a stainless steel foil with the film mass of (*a*) 165, (*b*) 373 and (*c*) 658 μg cm^{-2}.

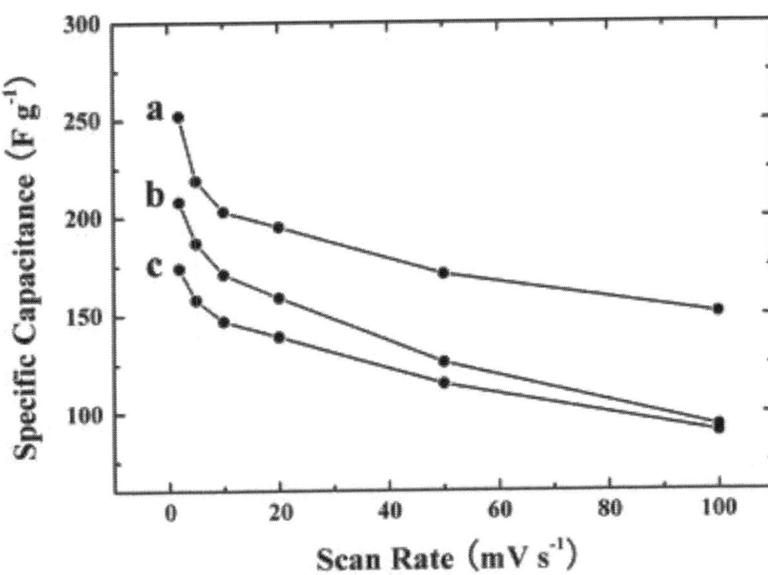

Figure 10. SC versus scan rate for the films deposited from the solution containing tiron on a stainless steel foil with the film mass of (*a*) 89, (*b*) 232 and (*c*) 416 μg cm^{-2}

The results of this investigation indicated that polypyrrole films were successfully deposited on stainless steel substrates from aqueous solutions of pyrrole, containing sodium salicylate and tiron additives. The films exhibited capacitive behavior and can be utilized for the fabrication of electrodes of ES using low-cost stainless steel current collectors. It is suggested that sodium salicylate and tiron adsorbed on the stainless steel substrates and passivated the surface of the substrates. The adsorption mechanism is related to chelating of surface metal ions [19]. The difference in the morphology and electrochemical behavior of the films prepared from solutions containing different additives can result from the properties of different anionic groups of salicylic acid and tiron.

CONCLUSIONS

Anodic electropolymerization method has been developed for the fabrication of polypyrrole films on stainless steel substrates for application in ES. The method is based on the use of sodium salicylate and tiron anionic additives, which were incorporated into the polymer to ensure the electrical neutrality of the growing film and passivated the surface of the stainless steel substrates. The deposition yield can be controlled by the variation of deposition time at a constant current density. SEM studies showed the formation of porous films with film thickness in the range of 0–3 μm. The film morphology is influenced by the additives. The films prepared using tiron additive showed lower particles size and improved uniformity compared to the films prepared using sodium salicylate. CV data for the films tested in the 0.5M Na_2SO_4 solutions showed capacitive behavior and a high SC in the voltage window of 0.9 V. The films deposited from pyrrole solutions containing tiron showed better capacitive behavior compared to those deposited from the solutions containing sodium salicylate additive. A highest SC of 254 F g^{-1} was observed for the sample with a specific mass of 89 μg cm^{-2} at a scan rate of 2 mV s^{-1}. The SC decreased with an increasing film thickness and scan rate. The results indicate that the polypyrrole films deposited on stainless steel substrates by anodic electropolymerization using tiron additive are promising electrode materials for ES.

ACKNOWLEDGMENTS

The authors gratefully acknowledge the financial support of the Natural Sciences and Engineering Research Council of Canada.

REFERENCES

1. P. Simon, Y. Gogotsi, Nat. Mater.7, 845 (2008)
2. H.T. Ham, Y.S. Choi, N. Jeong, I.J. Chung, Polymer46, 6308(2005)
3. L. Liu, Y. Zhao, Q. Zhou, H. Xu, C. Zhao, Z. Jiang, J. Solid State Electrochem.1, 32 (2006)
4. S. Lee, Y. Lee, M.-S. Cho, J.-D. Nam, J. Nanosci. Nanotechnol.8, 4722 (2008)
5. G.A. Snook, G.Z. Chen, D.J. Fray, M. Hughes, M. Shaffer,J. Electroanal. Chem.568, 135 (2004)
6. J. Wang, Y. Xu, X. Chen, X. Sun, Compos. Sci. Technol.67 ,2981 (2007)
7. A. Eftekhari, I. Ahmadi, Prog. Org. Coat.
8. 57, 371 (2006)
9. L.-Z. Fan, J. Maier, Electrochem. Commun.
10. 8, 937 (2006)
11. M.D. Ingram, H. Staesche, K.S. Ryder, Solid State Ionics 169, 51 (2004)
12. A.M.P. Hussain, A. Kumar, J. Power Sources 161, 1486 (2006)
13. S.K. Tripathi, A. Kumar, S.A. Hashmi, Solid State Ionics 177, 2979 (2006)
14. A. Izadi-Najafabadi, D.T.H. Tan, J.D. Madden, Synth. Met.152, 129 (2005)
15. B. Muthulakshmi, D. Kalpana, S. Pitchumani, N.G. Renganathan, J. Power Sources 158, 1533 (2006)
16. R.K. Sharma, A.C. Rastogi, S.B. Desu, Electrochem. Commun.10, 268 (2008)
17. Q. Zhang, X. Zhou, H. Yang, J. Power Sources 125, 141 (2004)
18. R. Ramya, M.V. Sangaranarayanan, J. Chem. Sci. 120, 25 (2008)
19. E. Hermelin, J. Petitjean, S. Aeiyach, J.C. Lacroix, P.C. Lacaze, J. Appl. Electrochem. 31, 905 (2001)
20. V.K. Gade, D.J. Shirale, P.D. Gaikwad, P.A. Savale, K.P. Kakde, H.J. Kharat, M.D. Shirsat, Int. J. Polym. Mater. 56, 167 (2007)
21. D.E. Tallman, C. Vang, G.G. Wallace, G.P. Bierwagen, J. Electrochem. Soc.149, C173 (2002

Chapter 5

Molecularly Imprinted Electropolymer for a Hexameric Heme Protein with Direct Electron Transfer and Peroxide Electrocatalysis

Lei Peng [1], Aysu Yarman [1], Katharina J. Jetzschmann [1], Jae-Hun Jeoung [2], Daniel Schad [2], Holger Dobbek [2], Ulla Wollenberger [1] and Frieder W. Scheller [1,3]

[1]*Institute of Biochemistry and Biology, Potsdam University, Karl-Liebknecht-Strasse 24-25, 14476 Potsdam, Germany*

[2]*Institute of Biology, Structural Biology/Biochemistry, Humboldt-Universität zu Berlin, Unter den Linden 6, 10099 Berlin, Germany*

[3]*Fraunhofer Institute for Cell Therapy and Immunology IZI-BB, Am Mühlenberg 13, 14476 Potsdam, Germany*

ABSTRACT

For the first time a molecularly imprinted polymer (MIP) with direct electron transfer (DET) and bioelectrocatalytic activity of the target protein is presented. Thin films of MIPs for the recognition of a hexameric tyrosine-coordinated heme protein (HTHP) have been prepared by electropolymerization of scopoletin after oriented assembly of HTHP on a self-assembled monolayer (SAM) of mercaptoundecanoic acid (MUA) on gold electrodes. Cavities which should resemble the shape and size of HTHP were formed by template removal. Rebinding of the target protein sums up the recognition by non-covalent interactions between the protein and the MIP with the electrostatic attraction of the protein by the SAM. HTHP bound to the MIP exhibits quasi-reversible DET which is reflected by a pair of well pronounced redox peaks in the cyclic voltammograms (CVs) with a formal potential of −184.4 ± 13.7 mV vs. Ag/AgCl (1 M KCl) at pH 8.0 and it was able to catalyze the cathodic reduction of peroxide. At saturation the MIP films show a 12-fold higher electroactive surface concentration of HTHP than the non-imprinted polymer (NIP).

Keywords: molecularly imprinted polymers; self-assembled monolayer; direct electron transfer; hydrogen peroxide; bioelectrocatalysis

1. INTRODUCTION

MIPs gain attention as artificial recognition elements both in scientific and industrial fields [1,2,3]. Molecular imprinting can be described as the formation of specific recognition sites in a material by polymerizing a monomer in the presence of a target molecule, the so-called template. During this process, functional monomers interact with complementary functional groups of the template and this complex is trapped in a polymeric matrix. After the template's removal by washing or extraction, cavities with specific size, shape, and functionalities are formed which recognize the template during the rebinding process [4].

Small molecules like sugars, steroids, drugs, and amino acid derivatives [5] have been successfully imprinted. Despite the importance of generating effective synthetic receptors for biomacromolecules less than 2% of the MIP literature is on macromolecular imprinting and only a few papers have described MIPs for enzymes [6,7,8,9,10,11,12,13,14,15]. Imprinting of biomacromolecular targets, especially proteins, is still a challenging task due to their large sizes, conformational flexibility, and instability in organic solvents [16].

Preparation of MIPs by electropolymerization circumvents several of these problems: electropolymerization makes it possible to optimize the film thickness in order to generate binding sites accessible to the target. Furthermore, the MIP formation proceeds directly on the transducer surface and can frequently be carried out in aqueous environments. These features are important for protein MIPs that require mild preparation procedures [17,18].

Recently we prepared an electrosynthesized MIP on top of a negatively charged SAM for the electron carrier protein cytochrome c (cyt c), where the oriented assembly of the protein on the negatively charged surfaces has been exploited to facilitate DET at the electrode [19]. Up to now bioelectrocatalysis for a protein MIP has not yet been reported in the literature. In this paper we describe a MIP with both DET and bioelectrocatalysis for a cosubstrate. This is demonstrated for the hexameric enzyme HTHP which exhibits an intrinsic peroxidatic activity. The bioelectrocatalytic enhancement of peroxide reduction gives evidence of both the productive binding inside the MIP cavities and the retained activity of the MIP-bound enzyme. Furthermore, the SAM/MIP architecture shows a preferential binding to its target in respect to cyt c, which otherwise strongly binds to the negatively charged (non-imprinted) SAM.

2. MATERIALS AND METHODS

2.1. Chemicals

Hydrogen peroxide (H_2O_2 30%), $[Ru(NH_3)_6]Cl_2$, 11-mercaptoundecanoic acid (MUA), cyt c (from equine heart, MW = 12,384 Da), and scopoletin (7-hydroxy-6-methoxycoumarin) were bought from Sigma-Aldrich (Steinheim, Germany). NADH was bought from Gerbu Biotechnic GmbH (Germany). All other reagents were of analytical grade and used without further purification.

HTHP, (pI 5.6), was prepared as described by Jeoung et al. [20]. It was expressed in Escherichia coli Rosetta (DE3) by induction with 1 µM isopropyl-ß-d-thiogalactopyranoside. Approximately 50 mg HTHP per liter of culture were purified to homogeneity as determined by a single band of approximately 7.9 kDa (for the monomeric form) on SDS-PAGE using standard protein purification protocols. Purified HTHP showed a R.Z. value of 0.26 (ratio of absorbance at Soret peak and 280 nm), which was increased to 2.8 by reconstitution of this preparation with equal molar hemin chloride. The final concentration of the purified hexameric HTHP was 1.3 mM in 50 mM Tris-HCl at pH 8 with 150 mM NaCl.

2.2. Preparation of Electrodes

Gold wire electrodes with a diameter of 0.5 mm and an active area of 0.161 cm^2 from Goodfellow, Germany, were boiled in 2 M KOH solution for 4 h and kept in concentrated HNO_3 for 10 min. After careful rinsing with Millipore water, they were stored in concentrated H_2SO_4 when not in use. Before every usage, the electrodes were washed with Millipore water and kept in concentrated HNO_3 for 10 min, then rinsed by Millipore water again in each successive step.

The cleaned electrodes were incubated in 5 mM MUA at least overnight at 4 °C. MUA was dissolved in 96 % ethanol and freshly prepared each time before modification. After being washed in Millipore water, a MUA modified gold wire electrode was immersed directly in 1.3 mM HTHP solution for 3 h at 4 °C to get the HTHP loaded electrode.

For the preparation of the HTHP-MIP polyscopoletin was deposited on the MUA covered Au electrode by electropolymerization from an aqueous solution of 0.5 mM scopoletin and 5 mM NaCl. A single potential pulse of 0.7 V for 5 s was followed by 0 V for 5 s. After formation of the HTHP-MIP, the modified electrodes were rinsed with water. The template protein, HTHP, was removed by incubating the HTHP-MIPs in 50 mM glycine-HCl (pH 2.2) on a shaker at 300 rpm for 1 h (25 °C). After template removal, the MIPs were rinsed with water and and could be stored in 2.5 mM phosphate buffer (PB) at pH 7 for one week. NIPs were prepared in the same manner but in absence of the protein and incubated in 50 mM glycine-HCl (pH 2.2) before the measurements.

2.3. Apparatus and Electrochemical Measurements

Electrochemical measurements were carried out with a PalmSens (Utrecht, Netherlands) electrochemical station. A three-electrode system with a working electrode, a Pt wire as the counter electrode and an Ag/AgCl (1 M KCl) as the reference electrode was used in all electrochemical experiments. Both CV and square wave voltammetry (SWV) were conducted in a 2 mL compartment of a custom-built reaction chamber with an adjustable magnetic stirring system. All experiments were performed at room temperature (25 °C) with exclusion of oxygen.

The DET of HTHP-MIPs was recorded by CV in 2.5 mM Tris buffer or PB at pH 8 from −0.5 to 0.2 V at different scan rates. The permeability of the MIP layer for the redox marker $[Ru(NH_3)_6]^{2+}$ after electropolymerization, template removal and rebinding was characterized by CVs between −500 and 0 mV in 100 mM NaCl at a scan rate of 100 mV/s. The concentration dependence of HTHP rebinding to the MIPs were performed after 1 h of incubation in HTHP-containing solutions in 2.5 mM PB at pH 7 by CV and SWV.

Bioelectrocatalytic reduction of H_2O_2 was investigated by CV in 2.5 mM Tris buffer at pH 8 (25 °C) with exclusion of oxygen. After each addition of H_2O_2 to the working buffer, CVs were recorded at a scan rate of 5 mV/s after 2 min of reaction time. NADH oxidation by HTHP MIPs was studied in the presence of 10 µM H_2O_2, aliquots of NADH were injected into the working cell and CVs were recorded at a scan rate of 5 mV/s after 2 min of reaction time.

3. RESULTS AND DISCUSSION

3.1. Preparation of the MIP-Modified Electrode, Template Removal and Rebinding

The MUA-modified electrode which is loaded with HTHP showed a pair of well pronounced peaks in the CV (Figure S1). For the electrochemical MIP preparation it was inserted in 0.5 mM scopoletin solution containing 5 mM NaCl. The low ionic strength was used in order to prevent the dissociation of HTHP from the SAM. The electrode was polarized for 5 s at 0.7 V followed by 5 s at 0 V. In this process scopoletin is polymerized and forms a network around the template HTHP (Figure 1). Scopoletin was chosen as the monomer [21] because in previous work we succeeded to prepare MIPs for cyt c [19] and the lectin concanavalin A (ConA) [22] by electropolymerizing it on top of a SAM.

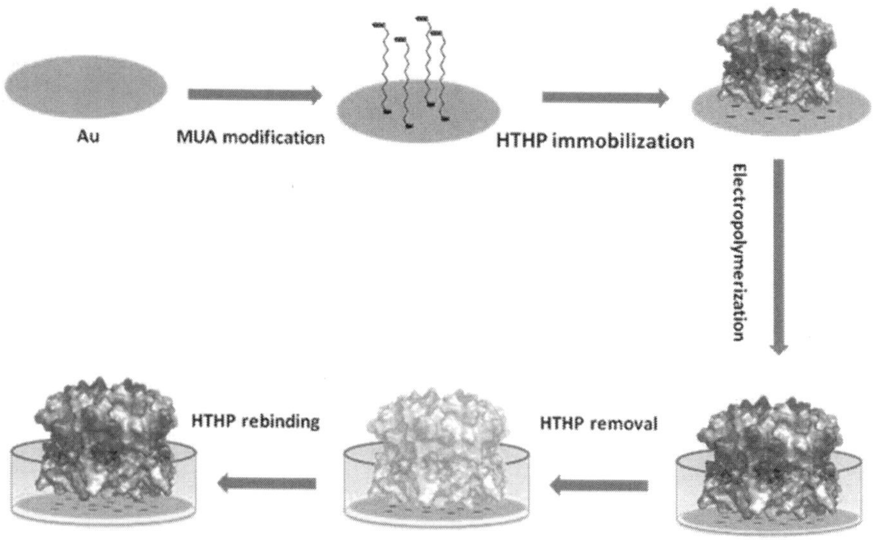

Figure 1. Schematic representation of the MIP preparation on a negatively charged thiol terminated SAM; Red: negatively charged region of HTHP, blue: positive region.

In order to characterize each step of the MIP-preparation, we studied the diffusive permeability of redox markers and the DET of the heme prosthetic groups by CVs. Whilst the redox peaks of the negatively charged ferricyanide are completely suppressed by the MUA-SAM, the peak currents of the positively charged $[Ru(NH_3)_6]^{2+}$ are almost 40 percent of the currents of the bare electrode (Figure 2b). Formation of the MIP layer by electropolymerization on top of the HTHP loaded MUA-SAM brought about a further decrease of the peaks of $[Ru(NH_3)_6]^{2+}$ by almost 55 percent (Figure 2c and Figure S2). Treatment of the MIP electrode with glycine-HCl resulted in a marked increase of the peak currents (Figure 2d and Figure S2) which indicates the formation of "diffusion pathways" for the small redox marker by the removal of the template. After incubation in HTHP containing solution (Figure 2e) the peak currents decreased to the values after electropolymerization. Parallel experiments showed that the peaks of the DET of the heme groups of

HTHP were decreased by electropolymerization. Treatment of the HTHP-MIP with glycine-HCl brought about an almost complete depression of the DET signal of HTHP (Figure S3). Rebinding of the target resulted in the reestablishment of the DET peaks. Both of the results with the redox marker, and for the DET, indicate that the target HTHP was effectively removed by the treatment with glycine-HCl and afterwards rebound from the HTHP containing solution.

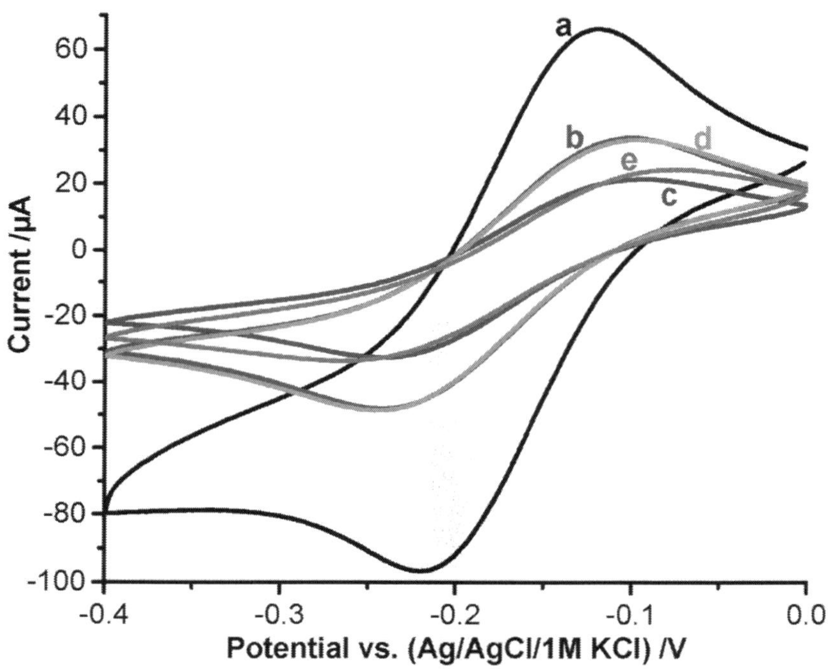

Figure 2. CVs of the redox marker $[Ru(NH_3)_6]^{2+}$ for the different steps of MIP preparation (5 mM $[Ru(NH_3)_6]^{2+}$ in 100 mM NaCl, pH 8, 100 mV/s): a—bare Au wire, b—after SAM-formation, c—after electropolymerization in presence of the template HTHP, d—after removal of HTHP, e—after rebinding in 1.3 mM HTHP solution for 1 h.

Under comparable conditions of electropolymerization, the thickness of the scopoletin-based MIP containing ConA was estimated by atomic force microscopy (AFM) and surface plasmon resonance (SPR) [22] to be 3.2 nm. This thickness of the polyscopoletin layer is comparable to the dimension of the

HTHP molecule (17.3 nm × 17.3 nm × 4.59 nm) [20] which may explain the effective template removal.

3.2. DET of HTHP Trapped in the MIP

HTHP from Silicibacter pomeroyi has a hexameric ring structure with a molecular mass of about 54 kDa. The six monomers are equivalent and contain one non-covalently bound heme in a hydrophobic pocket. The iron is coordinated by tyrosine in the proximal side while the distal side is encompassed by arginine [20]. The isoelectric point of HTHP is 5.6, so the overall charge at pH 8 should be negative. Theoretical calculations show that, on negatively charged surfaces, the HTHP disc binds to the surface via its neutrally charged side (Figure 1). In this "disc model", the distance between the six heme groups and the surfaces is shortest hence an efficient electron transfer could be expected [23].

HTHP adsorbed on MUA modified electrodes (HTHP-MUA-Au) displayed a pair of redox peaks with peak potentials at −165 and −251 mV. These redox peaks are typical for the Fe^{2+}/Fe^{3+} couple of heme containing proteins [24]. The appearance of only one peak pair and its shape is in accordance with an electrode reaction of a multi-redox center species with non- or weakly interacting centers [25].

After the formations of the MIP by electropolymerization on top of the HTHP loaded MUA-SAM the pair of the remaining peaks possessed a formal potential of −184.4 ± 13.7 mV vs. Ag/AgCl 1 M KCl at pH 8.0 (Figure 3). Compared to the value for HTHP at MUA modified electrode [23], it is only slightly positively shifted, which indicates that HTHP was not greatly altered by electropolymerization.

As shown in Figure 3, after rebinding the CVs of HTHP-MIPs showed almost symmetric peaks with equal heights for the reduction and the oxidation peaks which indicates that the surface concentration of electroactive HTHP in an oxidized and reduced state is very similar and the protein is not desorbed during the reduction-oxidation cycles. The peak separation ΔEp was 45 mV at scan rates of 100 $mV \cdot s^{-1}$. In the overall range of the scan rates, neither anodic nor cathodic peak currents increase linearly with increasing scan rates or with the square root of scan rates which means that neither a purely surface-controlled nor a purely diffusion-controlled electrode process took place.

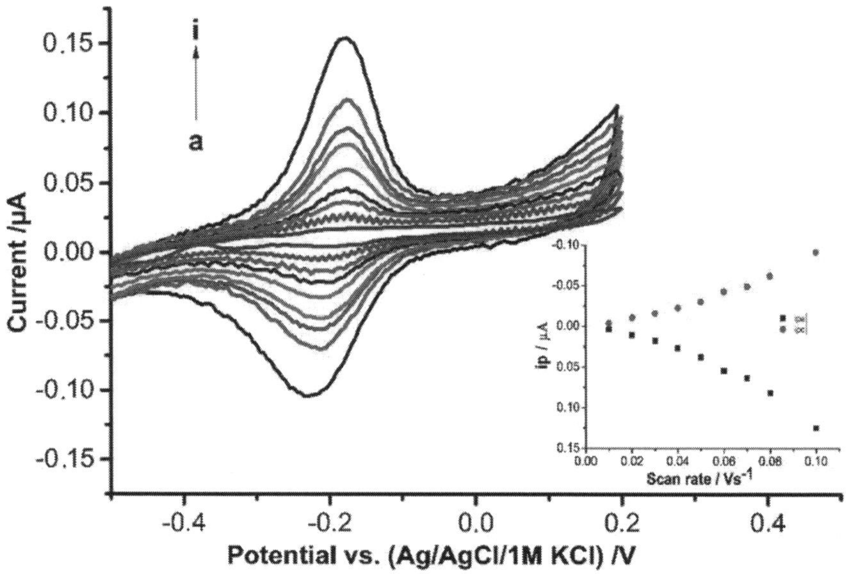

Figure 3. CVs at different scan rates for HTHP-MIP (after rebinding in 1.3 mM HTHP solution), a—i: 10, 20, 30, 40, 50, 60, 70, 80, 90, 100 mV/s in 2.5 mM Tris buffer, pH 8. Inset: Dependence of peak currents on scan rates.

3.3. Concentration Dependence of Rebinding of HTHP to MIPs and NIPs

Rebinding of HTHP to the MIPs and NIPs after template removal was investigated by SWV. After incubation of the MIPs in a series of HTHP solutions, the redox peaks of HTHP reappeared and the peak currents increased with increasing concentration of HTHP starting from approximately 30 μM and approached saturation above 100 μM (Figure 4). Current signal increased linearly with increasing concentration. On the other hand the peak currents for the NIPs are much smaller and almost constant within this concentration range. The signal for NIPs should be attributed to "non-specific" pores in the NIP film which allows the protein to reach the MUA-SAM.

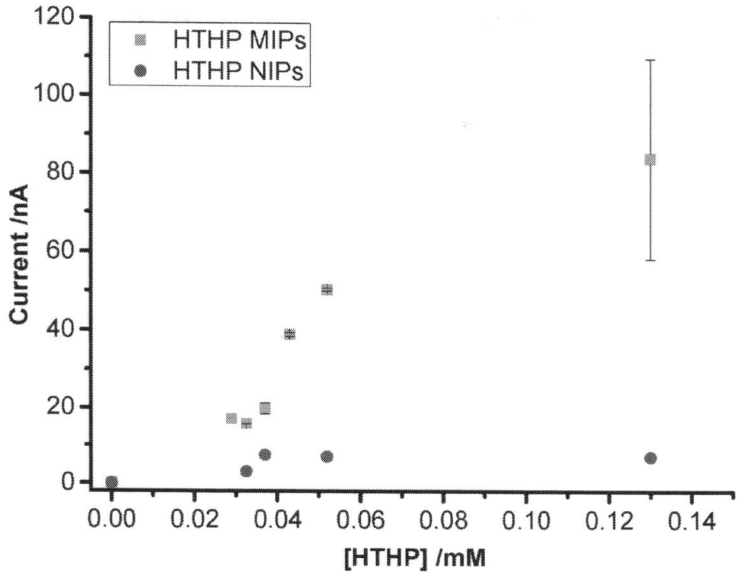

Figure 4. SWV peak currents for MIP- and NIP-modified electrodes after incubation in 2.5 mM K_2HPO_4–KH_2PO_4 pH 7 and different HTHP concentrations.

The evaluation of DET quantifies the "productively" rebound target molecules which might be different from the total amount of target bound to the MIP or NIP. In this respect, it is more specific than the measurement of the "total material bound" by quartz crystal balance or surface plasmon resonance or the very indirect measurement of the permeation of a redox marker by CV or impedance spectroscopy. The ratio of the DET signals at MIP and NIP—which represents a "functional" imprinting factor (IF)—was calculated to be 12 ± 3 at saturation. This clearly shows the higher efficiency of "productive" binding of the target protein to the imprinted film in relation to the NIP. Using the same polymer, the IF value for the cyt c-MIP on a MUA-SAM was 2 [19], and 8.6 for ConA on a mannose terminated SAM (Table 1) [22]. The lower value for the cyt c-MIP is caused by the relatively high signal for the DET at the NIP. Obviously, the harsh treatment with 1 M sulfuric acid which was applied for the removal of the target cyt c generated defects in the polymer layer. On the other hand, the hexameric HTHP was efficiently extracted from the MIP by incubation in glycine-HCl (pH 2.2) probably via the dissociation into subunits of the hexameric protein. This treatment did not affect the HTHP signal at the NIP. Table 1 gives an overview on the literature of electrochemically prepared protein-MIPs. It shows that the majority of papers use a redox marker for the detection of target binding. Our paper describes DET and bioelectrocatalysis of the target which is bound to the MIP. Among the few papers with DET of the target it is the first MIP for an oligomeric enzyme. The IF of the HTHP-MIP belongs to the highest values of protein-MIPs and indicates the remarkably higher electroactive surface concentration of HTHP for the MIP as compared with the NIP.

Table 1. Examples of protein MIPs prepared by electropolymerization.

Monomer	Template/Protein	IF	Detection of Binding	Ref.
Scopoletin	Cyt *c*-derived peptide	6	Fluorescence, SPR	[26]
Scopoletin	Cyt *c*	2	Fluorescence, CV (DET)	[19]
Scopoletin	ConA	8.6	QCM, SPS	[22]
Scopoletin	Ferritin	13	SPR, AFM	[27]
Scopoletin	HTHP	12	CV (DET, RM), SWV	This work
Pyrrole	*gp51*	9–10	PAD (RM)	[28]
Pyrrole	Bovine hemoglobin	-	DPV (RM), EIS, SEM	[29]
Pyrrole	Bovine hemoglobin	7.72	DPV (RM), EIS, SEM	[30]
3,4-Ethylenedioxythiophene/ Poly(styrenesulphonate)	Avidin	6.5	QCM, AFM	[31]
3,4-Ethylenedioxythiophene/ Poly(styrenesulphonate)	Avidin or Av-FITC	-	Fluorescence, SEM	[32]
o-Phenylenediamine	Troponin T	-	CV (RM), DPV, AFM	[33]
Phenol	Human ferritin/Human papillomavirus	-	EIS, DPV (RM), SEM	[34]
3-Aminophenylboronic acid	BSA	2	SPR, SEM	[35]
3,4-Propylenedioxythiophene carboxylate	Acetylcholinesterase	9.9	Amperometry of catalysis, AFM	[36]
2,2′-Bithiophene-5-carboxylic acid	Human serum albumin	26.8	DPV (RM), EIS, AFM	[37]

QCM: quartz crystal microbalance, SPS: surface plasmon spectroscopy, RM: redox marker, *gp51*: bovine leukemia virus glycoprotein *gp51*, PAD: pulsed amperometric detection, DPV: differential pulse voltammetry, EIS: electrochemical impedance spectroscopy, SEM: scanning electron microscopy, Av-FITC: avidin-fluorescein isothiocyanate, BSA: bovine serum albumin.

3.4. Binding of Cyt c to the HTHP-MIP

In order to check whether the MIP would preferentially bind its target HTHP, the binding of HTHP and the positively charged cyt c was studied on both MIP and MUA-SAM modified electrodes at low ionic strength. After incubation of the MIPs in 2.5 mM K_2HPO_4-KH_2PO_4 containing 32.5 µM HTHP, the SWV signal was well pronounced. It was almost 13 times higher than that of the MUA-modified gold electrode (Figure S4). On the other hand, the peak current for cyt c was almost 3.5 times smaller at the MIP/MUA electrode compared with that at the MUA-covered surface. These results show a preferential binding of the (negatively charged) HTHP to the MIP-covered SAM whilst the positively charged cyt c interacts more effectively with the negatively charged SAM.

The peak current for cyt c at the MIP-covered electrode is about 5 times smaller than the value for HTHP whilst at the bare MUA-SAM it is 8 times bigger than that for HTHP. The different influence of the SAM and the MIP layer

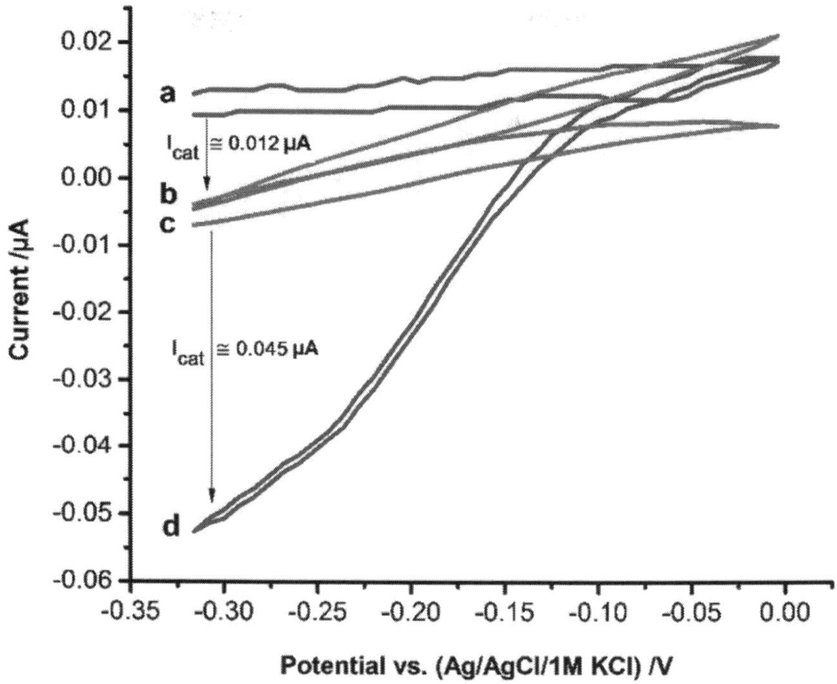

Figure 5. CVs of NIP (a, b) and HTHP loaded MIP (c, d) in absence (a, c) and in presence of 100 μM H_2O_2 (b, d) in 2.5 mM Tris buffer at pH 8, scan rate of 5 mV/s. MIP and NIP were incubated in 52 μM HTHP solution for 1 h before measurement.

may be caused by the fact that HTHP binds via the neutral side of the "disc" only weakly to the MUA [14] whilst cyt c is electrostatically bound via the positively charged lysine residues of the heme surrounding [38,39]. On the other hand, the MIP layer interacts favourably with its target HTHP as compared with cyt c. In spite of the smaller diameter of cyt c the target HTHP reaches the MUA-SAM more effectively which ensures the productive orientation for the DET.

3.5. Electrocatalysis of Hydrogen Peroxide Reduction

The bioelectrocatalytic reduction of H_2O_2 by HTHP MIPs was studied by incubating the electrodes in 52 μM HTHP solution for 1 h after the removal step. As shown in Figure 5, with curve d the catalytic current starts from around −0.1 V upon addition of hydrogen peroxide. Since no mediator is present, the reduction current should arise from the DET between the electrode and the heme protein [40,41]. However, the potential is 650 mV more negative than expected for the reduction of $Fe^{4+} = O^{-}$ center of Compound I [42] but in accordance with reports on heme protein-modified electrodes describing the peroxide reduction at the potential where the heme is in the reduced state [43,44]. The catalytic current increased linearly with increasing concentration of H_2O_2 in the range from 10 to 100 µM and reached saturation at 150 µM. This behavior is comparable with the effect of other heme proteins on the reduction of H_2O_2 [40,44,45]. NIPs which were not in contact with HTHP gave a H_2O_2 reduction current (Figure 5b) which is almost 3.8 times smaller than that of the HTHP MIPs.

Furthermore the peroxidatic activity of the HTHP-MIP towards NADH in the presence of H_2O_2 was studied. However, addition of 100 µM NADH did not cause a measurable change in the H_2O_2 reduction signal. This result might be explained by the low specific activity of 3 Units/mg of HTHP [20].

4. CONCLUSIONS

MIPs for the recognition of the hexameric heme protein HTHP with both DET and bioelectrocatalytic activity have been prepared by electropolymerization of scopoletin after oriented assembly of the target on a negatively charged SAM. Rebinding of the target to the MIP sums up the electrostatic attraction of the protein by the SAM with the shape recognition by non-covalent interactions with the MIP. This is reflected by the higher electroactive surface concentration of HTHP on the MIP covered SAM as compared with the SAM-modified electrode.

The evaluation of DET or bioelectrocatalysis quantifies the "productively" rebound target molecules. Therefore, it is more specific than the measurement of the total amount of bound material by quartz crystal balance or surface plasmon resonance or the indirect measurement of the permeation of a redox marker by cyclic voltammetry or impedance spectroscopy (Table 1). This principle is an efficient tool to develop MIPs for the almost 40 electroactive proteins described in the literature which will allow the regeneration of enzymes in sensors or fuel cells.

ACKNOWLEDGMENTS

This work has been part of UniCat, the Cluster of Excellence in the field of catalysis financially supported by the Deutsche Forschungsgemeinschaft (DFG) within the framework of the German Excellence Initiative (EXC 314).

REFERENCES

1. Wulff, G.; Sarhan, A. Use of polymers with enzyme-analogous structures for resolution of racemates. Angew. Chem. Int. Ed. **1972**, 11, 341–342.

2. Shea, K.J.; Thompson, E.A.; Pandey, S.D.; Beauchamp, P.S. Template synthesis of macromolecules—Synthesis and chemistry of functionalized macroporous polydivinylbenzene. J. Am. Chem. Soc. **1980**, 102, 3149–3155.

3. Arshady, R.; Mosbach, K. Synthesis of substrate-selective polymers by host-guest polymerization. Macromol. Chem. Phys. **1981**, 182, 687–692.

4. Ge, Y.; Turner, A.P.F. Too large to fit? Recent developments in macromolecular imprinting. Trends Biotechnol. **2008**, 26, 218–224.

5. Bowman, M.A.E.; Allender, C.J.; Brain, K.R.; Heard, C.M. A high-throughput screening technique employing molecularly imprinted polymers as biomimetic selectors. Methodol. Surv. Bioanal. Drugs **1998**, 25, 37–43.

6. Hayden, O.; Haderspöck, C.; Krassnig, S.; Chena, X.; Dickert, F.L. Surface imprinting strategies for the detection of trypsin. Analyst **2006**, 131, 1044–1050.

7. Bossi, A.; Piletsky, S.A.; Piletska, E.V.; Righetti, P.G.; Turner, A.P.F. Surface-grafted molecularly imprinted polymers for protein recognition. Anal. Chem. **2001**, 73, 5281–5286.

8. Chen, X.; Yang, Z.; Si, S. Potentiometric urea biosensor based on immobilization of urease onto molecularly imprinted TiO_2 film. J. Electroanal. Chem. **2009**, 635, 1–6.

9. Chen, Y.W.; Rick, J.; Chou, T.C. A systematic approach to forming micro-contact imprints of creatine kinase. Org. Biomol. Chem. **2009**, 7, 488–494.

10. Tan, C.J.; Tong, Y.W. The effect of protein structural conformation on nanoparticle molecular imprinting of ribonuclease a using miniemulsion polymerization. Langmuir **2007**, 23, 2722–2730.

11. Gao, R.; Zhang, L.; Hao, Y.; Cui, X.; Liu, D.; Zhang, M.; Tang, Y. Novel polydopamine imprinting layers coated magnetic carbon nanotubes for specific separation of lysozyme from egg white. Talanta **2015**, 144, 1125–1132.

12. Guerreiro, A.; Poma, A.; Karim, K.; Moczko, E.; Takarada, J.; de Vargas-Sansalvador, I.P.; Turner, N.; Piletska, E.; de Magalhães, C.S.; Glazova,

N.; et al. Influence of surface-imprinted nanoparticles on trypsin activity. Adv. Healthc Mater. **2014**, 3, 1426–1429.

13. Wang, S.; Ye, J.; Bie, Z.; Liu, Z. Affinity-tunable specific recognition of glycoproteins via boronate affinity-based controllable oriented surface imprinting. Chem. Sci. **2014**, 5, 1135–1140.

14. Kamon, Y.; Matsuura, R.; Kitayama, Y.; Ooya, T.; Takeuchi, T. Precisely controlled molecular imprinting of glutathione-s-transferase by orientated template immobilization using specific interaction with an anchored ligand on a gold substrate. Polym. Chem. **2014**, 5, 4764–4771.

15. Erdőssy, J.; Horváth, V.; Yarman, A.; Scheller, F.W.; Gyurcsányi, R.E. Electrosynthesized molecularly imprinted polymers for protein recognition. Trends Anal. Chem. **2016**.

16. Li, S.J.; Cao, S.S.; Whitcombe, M.J.; Piletsky, S.A. Size matters: Challenges in imprinting macromolecules. Prog. Polym. Sci. **2014**, 39, 145–163.

17. Sharma, P.S.; Pietrzyk-Le, A.; D'Souza, F.; Kutner, W. Electrochemically synthesized polymers in molecular imprinting for chemical sensing. Anal. Bioanal. Chem. **2012**, 402, 3177–3204.

18. Malitesta, C.; Mazzotta, E.; Picca, R.A.; Poma, A.; Chianella, I.; Piletsky, S.A. MIP sensors—The electrochemical approach. Anal. Bioanal. Chem. **2012**, 402, 1827–1846.

19. Bosserdt, M.; Gajovic-Eichelmann, N.; Scheller, F.W. Modulation of direct electron transfer of cytochrome c by use of a molecularly imprinted thin film. Anal. Bioanal. Chem. **2013**, 405, 6437–6444.

20. Jeoung, J.-H.; Pippig, D.A.; Martins, B.M.; Wagener, N.; Dobbek, H. HTHP: A novel class of hexameric, tyrosine-coordinated heme proteins. J. Mol. Biol. **2007**, 368, 1122–1131.

21. Gajovic-Eichelmann, N.; Ehrentreich-Forster, E.; Bier, F.F. Directed immobilization of nucleic acids at ultramicroelectrodes using a novel electro-deposited polymer. Biosens. Bioelectron. **2003**, 19, 417–422.

22. Dechtrirat, D.; Gajovic-Eichelmann, N.; Bier, F.F.; Scheller, F.W. Hybrid material for protein sensing based on electrosynthesized mip on a mannose terminated self-assembled monolayer. Adv. Funct. Mater. **2014**, 24, 2233–2239.

23. Peng, L.; Utesch, T.; Yarman, A.; Jeoung, J.H.; Steinborn, S.; Dobbek, H.; Mroginski, M.A.; Tanne, J.; Wollenberger, U.; Scheller, F.W. Surface-tuned electron transfer and electrocatalysis of a hexameric tyrosine-coordinated heme protein (HTHP). Chemisty **2015**, 21, 7596–7602.

24. Peng, L.; Wollenberger, U.; Kinne, M.; Hofrichter, M.; Ullrich, R.; Scheibner, K.; Fischer, A.; Scheller, F.W. Peroxygenase based sensor for aromatic compounds. Biosens. Bioelectron. **2010**, 26, 1432–1436.

25. Molina, A.; Serna, C.; López-Tenés, M.; Moreno, M. Theoretical background for the behavior of molecules containing multiple interacting or noninteracting redox centers in any multipotential step technique and cyclic voltammetry. J. Electroanal. Chem. **2005**, 576, 9–19.

26. Dechtrirat, D.; Jetzschmann, K.J.; Stocklein, W.F.M.; Scheller, F.W.; Gajovic-Eichelmann, N. Protein rebinding to a surface-confined imprint. Adv. Funct. Mater. **2012**, 22, 5231–5237.

27. Bosserdt, M.; Bognár, J.; Lautner, G.; Witt, J.; Köhler, K.; Gajovic-Eichelmann, N.; Yarman, A.; Wittstock, G.; Scheller, F.W.; Gyurcsányi, R.E. Microelectrospotting as a new method for electrosynthesis of surface-imprinted polymer microarrays for protein recognition. Biosens. Bioelectron. **2015**, 73, 123–129.

28. Ramanaviciene, A.; Ramanavicius, A. Molecularly imprinted polypyrrole-based synthetic receptor for direct detection of bovine leukemia virus glycoproteins. Biosens. Bioelectron. **2004**, 20, 1076–1082.

29. Kan, X.W.; Xing, Z.L.; Zhu, A.H.; Zhao, Z.; Xu, G.L.; Li, C.; Zhou, H. Molecularly imprinted polymers based electrochemical sensor for bovine hemoglobin recognition. Sens. Actuat B Chem. **2012**, 168, 395–401.

30. Li, L.; Yang, L.L.; Xing, Z.L.; Lu, X.J.; Kan, X.W. Surface molecularly imprinted polymers-based electrochemical sensor for bovine hemoglobin recognition. Analyst **2013**, 138, 6962–6968.

31. Bognar, J.; Szucs, J.; Dorko, Z.; Horvath, V.; Gyurcsanyi, R.E. Nanosphere lithography as a versatile method to generate surface-imprinted polymer films for selective protein recognition. Adv. Funct. Mater. **2013**, 23, 4703–4709.

32. Menaker, A.; Syritski, V.; Reut, J.; Opik, A.; Horvath, V.; Gyurcsanyi, R.E. Electrosynthesized surface-imprinted conducting polymer microrods for selective protein recognition. Adv. Mater. **2009**, 21, 2271–2275.

33. Karimian, N.; Turner, A.P.; Tiwari, A. Electrochemical evaluation of troponin T imprinted polymer receptor. Biosens. Bioelectron. **2014**, 59, 160–165.

34. Cai, D.; Ren, L.; Zhao, H.Z.; Xu, C.J.; Zhang, L.; Yu, Y.; Wang, H.Z.; Lan, Y.C.; Roberts, M.F.; Chuang, J.H.; et al. A molecular-imprint nanosensor for ultrasensitive detection of proteins. Nat. Nanotechnol. **2010**, 5, 597–601.

35. Wang, Y.; Wei, T.X. Surface plasmon resonance sensor chips for the recognition of bovine serum albumin via electropolymerized molecularly imprinted polymers. Chin. Chem. Lett. **2013**, 24, 813–816.

36. Jetzschmann, K.J.; Jágerszki, G.; Dechtrirat, D.; Yarman, A.; Gajovic-Eichelmann, N.; Gilsing, H.-D.; Schulz, B.; Gyurcsányi, R.E.; Scheller, F.W. Vectorially imprinted electropolymer for acetylcholinesterase. Adv. Funct. Mater. **2015**, 25, 5178–5183.

37. Cieplak, M.; Szwabinska, K.; Sosnowska, M.; Bikram, K.C.C.; Borowicz, P.; Noworyta, K.; D'Souza, F.; Kutner, W. Selective electrochemical sensing of human serum albumin by semi-covalent molecular imprinting. Biosens. Bioelectron. **2015**, 74, 960–966.

38. Song, S.; Clark, R.A.; Bowden, E.F.; Tarlov, M.J. Characterization of cytochrome c/alkanethiolate structures prepared by self-assembly on gold. J. Phys. Chem. **1993**, 97, 6564–6572.

39. Davis, K.L.; Drews, B.J.; Yue, H.; Waldeck, D.H.; Knorr, K.; Clark, R.A. Electron-transfer kinetics of covalently attached cytochrome c/SAM/Au electrode assemblies. J. Phys. Chem. C **2008**, 112, 6571–6576.

40. Csoregi, E.; Jonssonpettersson, G.; Gorton, L. Mediatorless electrocatalytic reduction of hydrogen-peroxide at graphite-electrodes chemically-modified with peroxidases. J. Biotechnol. **1993**, 30, 315–337.

41. Presnova, G.; Grigorenko, V.; Egorov, A.; Ruzgas, T.; Lindgren, A.; Gorton, L.; Börchers, T. Direct heterogeneous electron transfer of recombinant horseradish peroxidases on gold. Faraday Discuss. **2000**, 116, 281–289.

42. Jönsson, G.; Gorton, L. An electrochemical sensor for hydrogen-peroxide based on peroxidase adsorbed on a spectrographic graphite electrode. Electroanalysis **1989**, 1, 465–468.

43. Dunford, H.B. Oxidations of iron(II)/(III) by hydrogen peroxide: From aquo to enzyme. Coord. Chem. Rev. **2002**, 311, 233–234.

44. Yarman, A.; Nagel, T.; Gajovic-Eichelmann, N.; Fischer, A.; Wollenberger, U.; Scheller, F.W. Bioelectrocatalysis by microperoxidase-11 in a multilayer architecture of chitosan embedded gold nanoparticles. Electroanalysis **2011**, 23, 611–618.

45. Peng, L.; Wollenberger, U.; Hofrichter, M.; Ullrich, R.; Scheibner, K.; Scheller, F.W. Bioelectrocatalytic properties of Agrocybe aegerita peroxygenase. Electrochim. Acta **2010**, 55, 7809–7813.

Chapter 6

Highly Conformal Electrodeposition of Copolymer Electrolytes Into Titania Nanotubes For 3D Li-Ion Batteries

Nareerat Plylahan

Aix Marseille Université, CNRS, Chemistry of Materials Research Group LP3

UMR 7341, Marseille F-13288, France

ABSTRACT

The highly conformal electrodeposition of a copolymer electrolyte (PMMA-PEO) into self-organized titania nanotubes (TiO_2nt) is reported. The morphological analysis carried out by scanning electron microscopy and transmission electron microscopy evidenced the formation of a 3D nanostructure consisting of a copolymer-embedded TiO_2nt. The thickness of the copolymer layer can be accurately controlled by monitoring the electropolymerization parameters. X-ray photoelectron spectroscopy measurements confirmed that bis(trifluoromethanesulfone)imide salt was successfully incorporated into the copolymer electrolyte during the deposition process. These results are crucial to fabricate a 3D Li-ion power source at the micrometer scale using TiO_2nt as the negative electrode.

Keywords: Titania Nanotubes Electrodeposition Copolymer electrolyte Microbatteries

BACKGROUND

Nowadays, microbatteries are in demand as power source to drive small devices such as smartcards, medical implants, sensors, etc. To date, the electrochemical performances of these all-solid-state batteries are limited because planar thin films are employed as electrode and electrolyte materials. In general, the total thickness of the stacking films is below 15 μm, and the resulting battery reveals relatively low power and energy densities. In order to ensure significant advances for extended applications, it is crucial to improve the electrochemical performances by investigating new materials and manufacturing processes. In this context, the large specific area offered by nano-architectured electrodes represents a promising alternative to improve the general performances of these micro-power sources [1].

Particularly, better rate capability, capacity, and cycling behavior have been observed for self-organized nanostructures such as titania nanotubes (TiO_2nt) [2–9]. However, when targeting 3D microbatteries, the conventional top-down approach to deposit solid electrolyte (e.g., lithium phosphorous oxynitride) [10–12] is not really suitable due to the accumulation of the electrolyte at the top of the nanotubes [13]. Certainly, with this accumulation of electrolyte, the 3D paradigm of microbatteries cannot be realized. Thus, investigating the deposition of polymer electrolytes into nanostructures by electrochemical techniques is a convenient way to ensure the desired filling of the nanostructures [14]. Indeed, electropolymerization is particularly powerful to control the deposition of different polymers into various porous materials [15–19]. Very recently, the use of electrodeposition to fill TiO_2nt with a layer of poly(methyl methacrylate)-polyethylene oxide, i.e., PMMA-$(PEO)_{475}$ has been reported [13, 20]. We have demonstrated that this simple bottom-up approach is adequate to deposit a homogeneous copolymer layer into titania nanotubes while improving the electrochemical performance. However, conformal coating of the nanotubes by the polymer electrolyte is required to design a 3D microbattery. In this work, it is reported that the conformal deposition of a PMMA-PEO electrolyte into self-organized TiO_2nt can be obtained by controlling the electrodeposition parameters. The morphology and the chemical composition of the resulting copolymer-embedded TiO_2nt materials are characterized by scanning electron microscopy (SEM) and transmission electron microscopy (TEM). The incorporation of lithium bis(trifluoromethanesulfone)imide, so-called LiTFSI salt, into the electrodeposited polymer is studied by X-ray photoelectron spectroscopy (XPS).

METHODS

Synthesis of self-organized TiO_2nt has been widely reported for a wide range of applications [21–24]. In the present work, TiO_2nt layers were produced by the electrochemical anodization of Ti foils using the Modulab potentiostat from Solartron Analytical (Hampshire, UK). Before the anodization, Ti foils with the 99.6+ % purity and 0.125-mm thickness were cut into pieces with the desired dimensions and sonicated sequentially during 10 min in acetone, propanol, and methanol sequentially. After that, the foils were rinsed with deionized water and dried with compressed air. The anodization process was carried out in an electrochemical cell containing a solution of 1 M H_3PO_4, 1 M NaOH, and 0.4 wt.% of HF. The setup consisted of Ti foil as the working electrode, a Pt grid as the counter electrode, and a Hg/Hg_2SO_4, K_2SO_4 (saturated) (E=0.64V vs NHE) reference electrode. A constant voltage of 20 V was applied during 2 h. The material was rinsed with deionized water and dried with compressed air immediately after the anodization process.

Then, an aqueous electrolyte containing 0.035 M LiTFSI was introduced into the cell and purged with N_2 gas for 10 min before adding 4 g of the MMA-$(PEO)_{475}$ monomer provided by Sigma Aldrich (St. Louis, MO, USA). It can be noticed that no initiator was added into the solution. The copolymer-embedded TiO_2nt was obtained by cyclic voltammetry (CV) using the as-prepared TiO_2nt layers as the working electrode and a Pt grid as the counter electrode. The CV curves were carried out in the potential window ranging from −0.4 to −2.5 V vs Hg/Hg_2SO_4, K_2SO_4 (saturated) with the scan rate of 25 mV/s. The number of cycles was varied from 1 to 10 in order to observe the influence of cycle number on the polymer electrolyte layers. After electropolymerization, the samples were dried at room temperature to evaporate part of the residual water. The morphology of the copolymer-embedded TiO_2nt was studied by SEM and TEM analyses using a JEOL 6320 F SEM and a JEOL 2010 F TEM (JEOL Ltd., Akishima, Tokyo, Japan). XPS measurements were carried out with a Kratos Axis Ultra spectrometer (Kratos Analytical Ltd., Manchester, UK), using focused monochromated Al Kα radiation (hν = 1,486.6 eV). The XPS spectrometer was directly connected to an argon dry box through a transfer chamber to avoid moisture/air exposure of the samples. The analyzed area of each sample was 300 µm × 700 µm. Peaks were recorded with constant pass energy of 20 eV. The pressure in the analysis chamber was around 5×10^{-8} Pa. Short acquisition time spectra were recorded before and after each normal experiment to check that the samples did not suffer from degradation under the X-ray beam during measurements. Peak assignments were made with respect to experimental reference compounds, namely bulk anatase and/or rutile TiO_2. The binding energy scale was calibrated from hydrocarbon contamination using the C 1 s peak at 285.0 eV. Core peaks were analyzed using a non-linear Shirley-type background. The peak positions and areas were optimized by a weighted least-square fitting method using 70% Gaussian and 30% Lorentzian line shapes. Quantification was performed on the basis of Scofield's relative sensitivity factors.

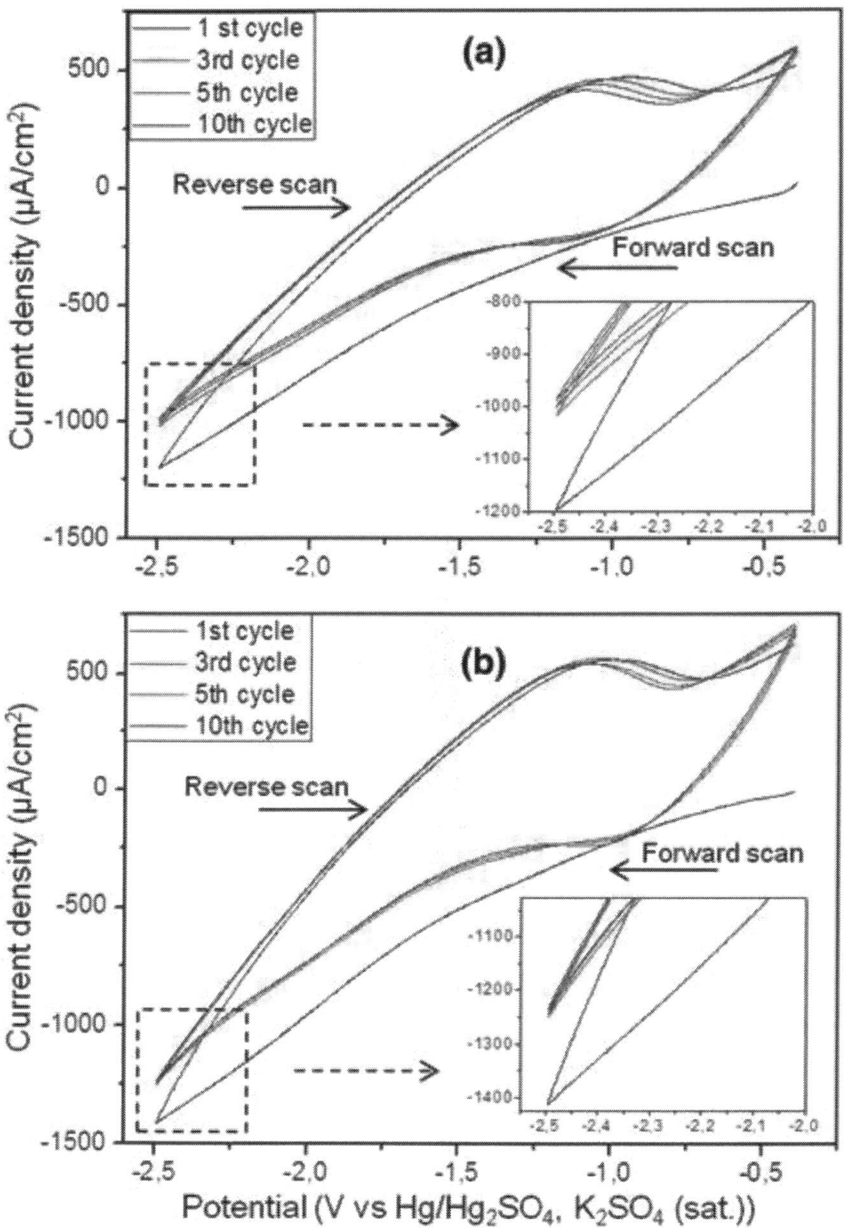

Figure 1. Cyclic voltammograms of TiO 2 nt electrode. In (**a**) 0.035 M LiTFSI and (**b**) in 0.035 M LiTFSI + 4 g MMA-PEO at the 1st, 3rd, 5th, and 10th cycles. The electropolymerization was carried out between −0.4 and −2.5 V vs Hg/Hg_2SO_4, K_2SO_4 (saturated) with the scan rate of 25 mV/s.

RESULTS AND DISCUSSION

The electropolymerization of MMA-$(PEO)_{475}$ monomer was carried out by cyclic voltammetry (CV) given in (Figure 1a). For comparison, the CV curves recorded in monomer-free electrolyte are given in Figure 1b. It is important to note that the monomer is not electrochemically active, and the polymerization process is expected to be initiated by free hydrogen radicals produced at applied cathodic potentials higher than −1 V vs Hg/Hg_2SO_4, K_2SO_4 (saturated) [13, 20, 25]. The similar shape of the CVs recorded in the two different electrolytes can be explained by the reactions involving the Ti^{4+}/Ti^{3+} and H^+/H_2 redox couples. However, variations of the current densities obtained in the monomer-free electrolyte are almost independent of cycle number, suggesting that all electrochemical reactions are reversible without modification of the surface. In contrast, the cathodic current density decreases slightly with cycle number when the monomer is present. This effect on current density can be attributed to the thin copolymer layers successively deposited onto the titania nanotube walls and, consequently, acting as electrical insulators.

The SEM images of the as-formed TiO_2nt and the copolymer-embedded TiO_2nt with different numbers of CV cycles are displayed in Figure 2. The evolution in thickness of the copolymer layer can clearly be observed. The SEM image of the as-formed TiO_2nt shown in Figure 2a reveals very thin tube walls (around 9 nm) and presents some degree of perforation at the top. As the cycles of CV increase from 0 to 1, 5, and up to 10 according to Figure 2b,c,d, the average thickness of the nanotube walls increases from 9 to 13, 17, and 18 nm (see Table 1), respectively, while the inner diameters of the tubes and spaces between the tubes decrease from 90 to 84, 72 and 70 nm (Table 1), respectively. It can also be noticed that the perforation in the tube walls has clearly disappeared. These results confirm that the growth of a copolymer layer occurred on the inside and outside of each nanotube wall.

The deposition of the copolymer was further confirmed by examination of the cross-sectional SEM images given in Figure 3. Although the thin layer of copolymer is not obvious in cross-section after the first cycle (Figure 3a), it can easily be discerned along the nanotubes after 10 cycles of CV (Figure 3b). The inter-tube spaces are filled with the copolymer, and the nanotube walls become significantly thicker as compared to the sample obtained after 1 cycle of CV. It can also be noticed in these cross-sectional images (Figure 3a,b) that the perforations at the top of the tube are no longer visible owing to the filling and covering with the copolymer.

In order to verify the conformal deposition of the copolymer, TEM imaging was performed on a single nanotube. The TEM images of the TiO_2nt after 5 cycles of CV are displayed in Figure 4. From Figure 4a, it can clearly be observed that a homogeneous copolymer layer of 6 nm in thickness has been deposited onto the inner and outer walls of the nanotube. This conformal coating along the nanotube wall can be confirmed by examining the bottom part of the tube (Figure 4b). A thicker deposit is observed at the bottom of the tube, suggesting that the copolymer initially grows from the bottom to the top of the tube.

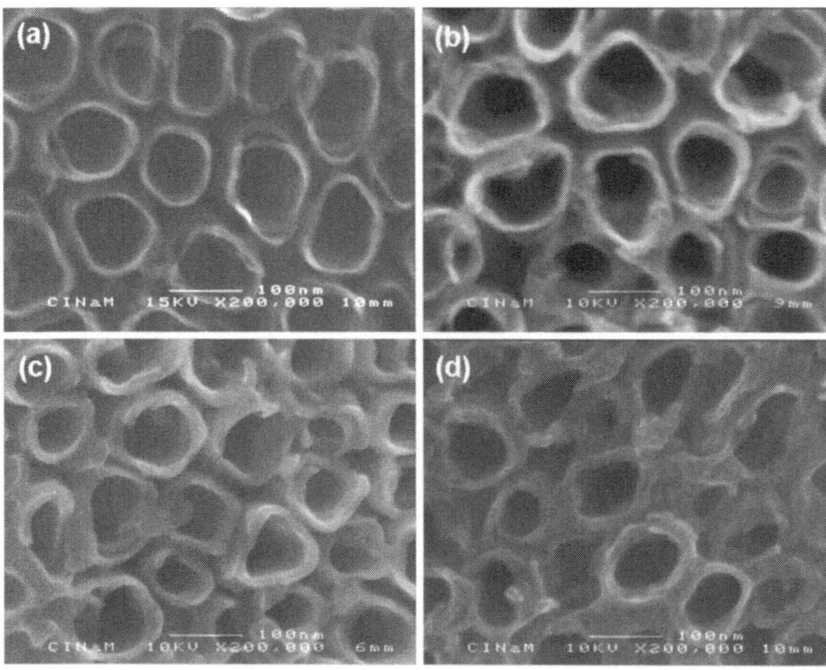

Figure 2: SEM images. (**a**) As-formed TiO_2nt and (**b**) 1 cycle, (**c**) 5 cycles, and (**d**) 10 cycles copolymer-embedded TiO2nt.

Table 1. Thickness of nanotube walls and inner diameters of tubes.

Number of cycles	**Thickness of nanotube walls (nm)**	**Inner diameter of nanotubes (nm)**
0	9	90
1	13	84
5	17	72
10	18	70

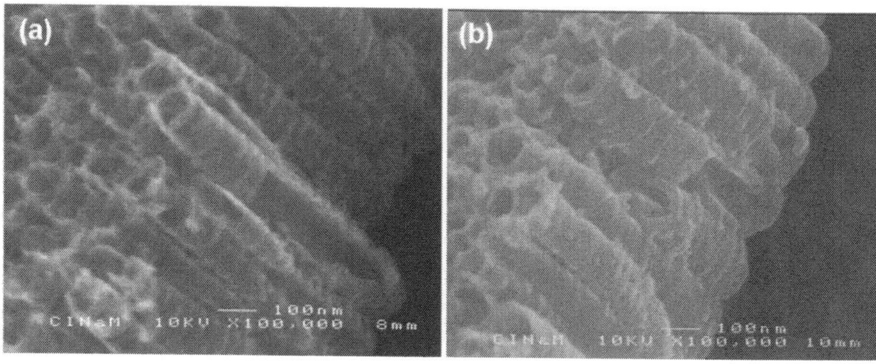

Figure 3: Cross-sectional images of the copolymer-embedded TiO $_2$ nt after (a) 1 cycle and (b) 10 cycles.

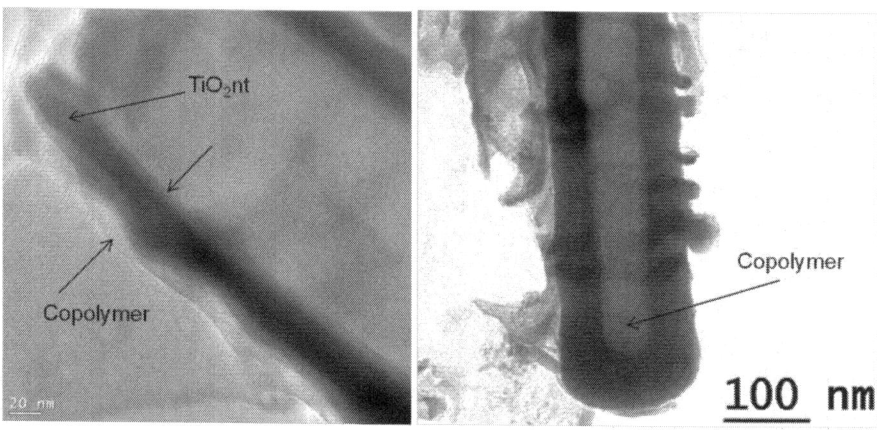

Figure 4: TEM images of copolymer-embedded TiO $_2$ nt after 5 cycles of CV. **(**a**)** Deposition of a homogeneous copolymer layer onto the inner and outer walls of the nanotube and (b**)** examination of the bottom part of the tube.

In order to confirm the presence of the copolymer and lithium salt (LiTFSI) in the self-organized TiO_2nt structure, XPS measurement was performed. The XPS spectra of C 1 s and O 1 s binding energies of TiO_2nt and copolymer-embedded TiO_2nt are displayed in Figure 5. Concerning the C 1 s core peak (Figure 5a), the binding energy of C-H and C-O is clearly different around 285 and 286.5 eV. For copolymer-coated TiO_2nt, the main contribution of this peak consists of a component located at 286.7 eV. This component is associated to the presence of oligomeric species of PEO $(-CH_2-CH_2-O-)_n$ for which all carbon atoms are in a one-oxygen environment [26]. A sole peak at 292.8 eV can be assigned to CF_3-

like carbon atoms in LiTFSI ($LiN(SO_2CF_3)_2$) as previously observed by Leroy et al. [27] in agreement with Ensling et al. [28]. The C 1s peak for TiO_2nt could be decomposed into three components respectively assigned to C-C (285.0 eV), CO (286.5 eV), and O-C = O (288.8 eV). For the O 1s spectra (Figure 5b), two main peaks of as-formed TiO_2 at 530.2 (Ti-O) and 531.5 eV (O-H hydroxyl) are absent in the copolymer-embedded TiO_2nt sample. The absence of these peaks results from the coverage of TiO_2nt with the copolymer. The broad peak of the copolymer-embedded TiO_2nt sample may be assigned to O in PEO and LiTFSI.

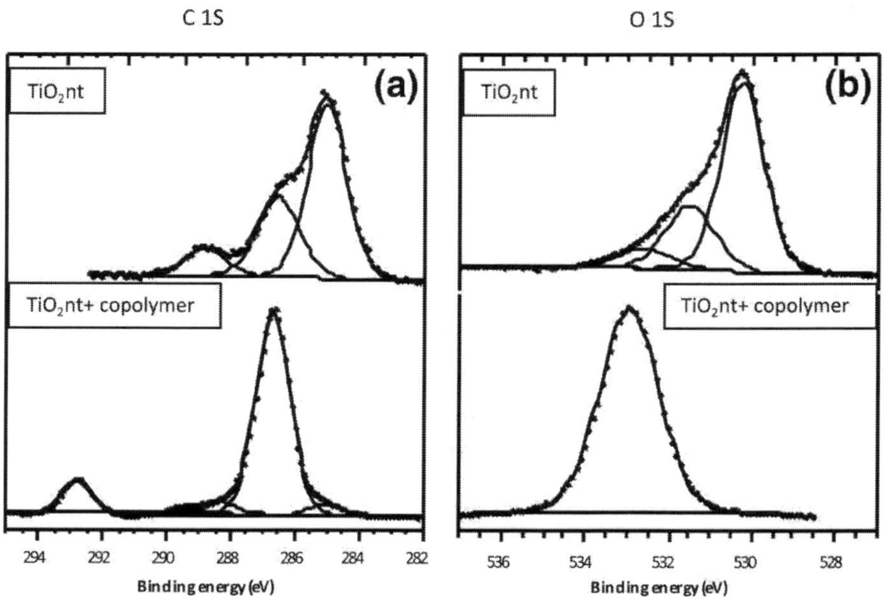

Figure 5. XPS spectra. **(a)** C 1s and **(b)** O 1s regions for as-formed TiO_2nt and copolymer-embedded TiO_2nt.

The presence of LiTFSI is confirmed by the XPS spectra of the copolymer-embedded sample at S 2p, F 1s and N 1s binding energies as shown in Figure 6. According to Leroy et al. [27], the XPS core peak characterization of LiTFSI was achieved. In their work, the binding energy of N 1s spectrum is located at 399.6 eV (399.4 eV in this work). The F 1s, Li 1s, and S $2p_{3/2}$ core peaks appear respectively at 688.6, 56.6, and 169.0 eV (688.7, 56.7, 168.8 eV in the present study). Concerning C 1s spectrum, a sole peak at 293.0 eV can be observed for LiTFSI ($LiN(SO_2CF_3)_2$), assigned to CF_3-like carbon atoms. Here, the C 1s core peak corresponding to C-F_3-like carbon atoms is located at 292.8 eV. In conclusion, our results are in perfect agreement with those previously published. We can, thus, conclude to the evident presence of LiTFSI in the copolymer-embedded sample.

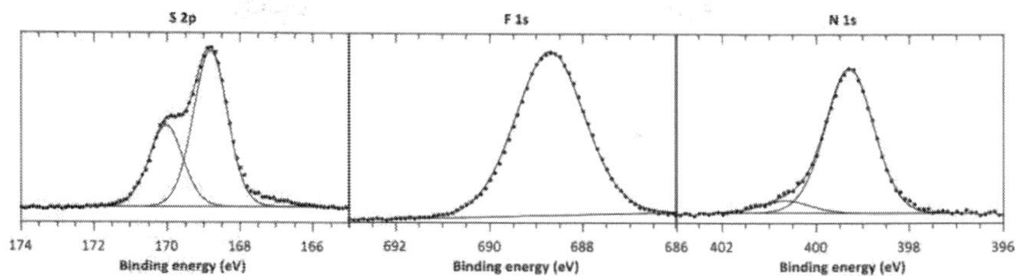

Figure 6: XPS spectra of S 2p, F 1 s, and N 1 s regions of copolymer-embedded TiO_2 nt.

CONCLUSIONS

The highly conformal electrodeposition of copolymer electrolyte has been successfully achieved on titania nanotubes. It is demonstrated that control of the electropolymerization parameters allows to homogeneously cover the nanostructures without closing the tubes. By this technique, the copolymer-embedded titania nanotubes retain the 3D structure which is advantageous for the further fabrication on high-performance 3D microbatteries.

ACKNOWLEDGMENTS

We acknowledge the French Ministry of Education, C'Nano PACA, Région PACA, and ANR JCJC number 2010 910 01 for financial support. We would like to thank Serge Nitsche and Damien Chaudanson from the electron microscopy service of CINaM Laboratory (UMR 7325) for their assistance in obtaining SEM images.

REFERENCES

1. Roberts M, Johns P, Owen J, Brandell D, Edstrom K, Enany GE, Guery C, Golodnitsky D, Lacey M, Lecoeur C, Mazor H, Peled E, Perre E, Shaijumon MM, Simon P, Taberna PL: 3D lithium ion batteries-from fundamentals to fabrication. J Mater Chem 2011, 21: 9876–9890.
2. Ortiz GF, Hanzu I, Knauth P, Lavela P, Tirado JL, Djenizian T: Nanocomposite electrode for Li-ion microbatteries based on SnO on nanotubular titania matrix. Electrochem Solid-State Lett 2009, 12(9):A186-A189.
3. Ortiz GF, Hanzu I, Knauth P, Lavela P, Tirado JL, Djenizian T: TiO2 nanotubes manufactured by anodization of Ti thin films for on-chip Li-ion 2D microbatteries. Electrochim Acta 2009, 54: 4262–4268.

4. Ortiz GF, Hanzu I, Lavela P, Knauth P, Tirado JL, Djenizian T: Nanoarchitectured TiO2/SnO: a future negative electrode for high power density Li-ion microbatteries? Chem Mater 2010, 22: 1926–1932.
5. Ortiz GF, Hanzu I, Lavela P, Tirado JL, Knauth P, Djenizian T: A novel architectured negative electrode based on titania nanotube and iron oxide nanowire composites for Li-ion microbatteries. J Mater Chem 2010, 20: 4041–4047.
6. Ortiz GF, Hanzu I, Lavela P, Knauth P, Djenizian T, Tirado JL: Novel fabrication technologies of 1D TiO2 nanotubes, vertical tin and iron-based nanowires for Li-ion microbatteries. Int J Nanotechnology 2012, 9: 260–294.
7. Djenizian T, Hanzu I, Premchand YD, Vacandio F, Knauth P: Electrochemical fabrication of Sn nanowires on titania nanotube guide layers. Nanotechnology 2008, 19: 205601–205605.
8. Du G, Guo Z, Zhang P, Li Y, Chen M, Wexler D, Liu H: SnO2 nanocrystals on self-organized TiO2 nanotube array as three-dimensional electrode for lithium ion microbatteries. J Mater Chem 2010, 20: 5689–5694.
9. Choi MG, Lee YG, Song SW, Kim KM: Lithium-ion battery anode properties of TiO2 nanotubes prepared by the hydrothermal synthesis of mixed (anatase and rutile) particles. Electrochim Acta 2010, 55: 5975–5983.
10. Bates JB, Dudney NJ, Neudecker BJ, Hart FX, Jun HP, Hackney SA: Thin-film lithium and lithium-ion batteries. J Electrochem Soc 2000, 147: 59–70.
11. Bates JB, Dudney NJ, Gruzalski GR, Zuhr RA, Choudhury A, Luck CF, Robertson JD: Electrical properties of amorphous lithium electrolyte thin films. Solid State Ionics 1992, 53: 647–654.

Chapter 7

Enzyme Immobilization Strategies and Electropolymerization Conditions to Control Sensitivity and Selectivity Parameters of a Polymer-Enzyme Composite Glucose Biosensor

Sharon A. Rothwell, Sarah J. Killoran and Robert D. O'Neill

UCD School of Chemistry and Chemical Biology, University College Dublin, Belfield, Dublin 4, Ireland

ABSTRACT

In an ongoing programme to develop characterization strategies relevant to biosensors for in-vivo monitoring, glucose biosensors were fabricated by immobilizing the enzyme glucose oxidase (GOx) on 125 μm diameter Pt cylinder wire electrodes (Pt_C), using three different methods: before, after or during the amperometric electrosynthesis of poly(ortho-phenylenediamine), PoPD, which also served as a permselective membrane. These electrodes were calibrated with H_2O_2 (the biosensor enzyme signal molecule), glucose, and the archetypal interference compound ascorbic acid (AA) to determine the relevant polymer permeabilities and the apparent Michaelis-Menten parameters for glucose. A number of selectivity parameters were used to identify the most successful design in terms of the balance between substrate sensitivity and interference blocking. For biosensors electrosynthesized in neutral buffer under the present conditions, entrapment of the GOx within the PoPD layer produced the design (Pt_C/PoPD-GOx) with the highest linear sensitivity to glucose (5.0 ± 0.4 μA cm^{-2} mM^{-1}), good linear range ($K_M = 16 \pm 2$ mM) and response time (< 2 s), and the greatest AA blocking (99.8% for 1 mM AA). Further optimization showed that fabrication of Pt_C/PoPD-GOx in the absence of added background electrolyte (i.e., electropolymerization in unbuffered enzyme-monomer solution) enhanced glucose selectivity 3-fold for this one-pot fabrication protocol which provided AA-rejection levels at least equal to recent multi-step polymer bilayer biosensor designs. Interestingly, the presence of enzyme protein in the polymer layer had opposite effects on permselectivity for low and high concentrations of AA, emphasizing the value of studying the concentration dependence of interference effects which is rarely reported in the literature.

Keywords: Hydrogen peroxide; polyphenylenediamine; amperometry; enzyme-modified electrode; ascorbic acid interference; brain monitoring

1. INTRODUCTION

The application of biosensors as analytical tools is a growing research topic in areas such as environmental surveillance, batch food analysis and clinical monitoring, and is beginning to impact on quality-of-life issues [1–5]. The choice of biosensor design for a particular application should be governed by diverse factors, including: the chemical nature of the analytical medium (e.g., lipophilic versus hydrophilic); the sample size (e.g., intracellular and extracellular monitoring versus batch analysis); the time resolution and recording duration required; and the concentration of the target analyte relative to the corresponding interference compounds for the chosen technique (electrochemical, optical, gravimetric, tonometric, thermal, magnetoelastic, etc.) [6–8]. For in-vivo monitoring in the brain during behavior, implantable biosensors showing good biocompatibility, sensitivity, selectivity and stability in this strongly lipophilic environment are needed, and amperometric enzyme-based devices incorporating a permselective polymer have been applied successfully in many neurochemical studies [9–17].

Poly-phenylenediamines (PPDs) electrosynthesized from one of the three monomer isomers have found widespread use as a biosensor permselectivity barrier [18–21], although poly(ortho-phenylenediamine), PoPD, may be superior for long-term in-vivo monitoring [22]. A variety of immobilization methods for oxidase enzymes (EOx) have also been described for PPD-based biosensors, with three approaches commonly used: enzyme deposited before the PPD layer, EOx/PoPD [23–26], enzyme immobilized over PPD, PPD/EOx [23,27–29] and enzyme co-immobilized from the monomer solution, PPD-EOx [30–32].

The amperometric enzyme-based biosensors used in this work were first generation devices which involve monitoring the formation of hydrogen peroxide, HP [33]. The first two reactions [Equations (1) and (2)] represent the enzyme (glucose oxidase, GOx) redox reactions, where FAD is the oxidized form of the prosthetic group, flavin adenine dinucleotide:

$$\beta\text{-D-glucose} + \text{GOx/FAD} \rightarrow \text{D-glucono-}\delta\text{-lactone} + \text{GOx/FADH}_2 \quad (1)$$

$$\text{GOx/FADH}_2 + O_2 \rightarrow \text{GOx/FAD} + H_2O_2 \quad (2)$$

$$H_2O_2 \rightarrow O_2 + 2H^+ + 2e^- \quad (3)$$

The H_2O_2 produced in Equation (2) can be oxidized, usually amperometrically, either directly on the electrode surface at relatively high applied potentials [Equation (3)] [33], or catalytically at lower potentials. However, even when significantly lower applied overpotentials can be used for H_2O_2 detection [34], interference by the ubiquitous biological reducing agent, ascorbic acid (AA), can persist because of its high concentration in most biological media and very low redox potential [35,36], and the use of redox-mediated HRP-based biosensors may suffer from indirect AA interference because HRP has been reported to catalyze the reaction between AA and H_2O_2 [37]. The incorporation of a permselective layer minimizes interference problems, and PoPD fulfils this function in many laboratories, blocking AA and other interference species well (dopamine, DOPAC, uric acid, etc.) while showing excellent permeability to H_2O_2 [10,11,18,25,26,31,38–40].

Recently, a number of new aspects to the problem of interference at PoPD-based biosensors have been identified. First, the permselectivity can be undermined for biosensors with large values of Pt-insulation "edge density", such as microdisks [41]. Briefly, PoPD deposited near the electrode insulation is not as effective at blocking interference. Second, the incorporation of enzyme in the PoPD can decrease its blocking ability [41]. Third, electrosynthesis of enzyme-free PoPD in the absence of added background electrolyte can improve its permselective properties, apparently due to fewer ions being trapped in the polymer matrix [42]. Taking cognisance of these new findings, here we chose an implantable, low edge-density geometry (narrow Pt-Ir cylinders) as electrode substrate, and investigate the effects of different enzyme immobilization methods and electropolymerization conditions on the enzyme-kinetic and permeability parameters. Important aims of this study were to understand more fully factors affecting the characteristics of PoPD-based polymer-enzyme composite devices, and to determine whether optimizing the fabrication conditions of these single-polymer-layer biosensors could provide interference-rejection characteristics comparable to more complex sensing layers, such as those formed from multiple electrosynthesis and over-oxidation steps [43].

2. EXPERIMENTAL SECTION

2.1. Chemicals and Solutions

The enzyme glucose oxidase (GOx) from Aspergillus niger (180,200 U·g^{-1}; EC 1.1.3.4, type VII-S) was obtained from Sigma-Aldrich, as were ortho-phenylenediamine (oPD), α-d-(+)-glucose, ascorbic acid (AA), hydrogen peroxide (HP, 30% w/w aqueous solution) and potassium chloride. All reagents were used as supplied. All experimental calibrations were carried out in PBS (pH 7.4) prepared by adding NaCl (Sigma, 150 mM), NaH_2PO_4 (Fluka, 40 mM) and NaOH (Fluka, 40 mM) to distilled water, bubbled with N_2 for 15 min, and stored at 4 °C. Solutions of monomer, oPD, were prepared in 25 mL of PBS, unless stated otherwise, and sonicated at room temperature until dissolved. A stock solution of 1 M glucose was prepared in distilled water and left for 24 h at room temperature to allow equilibration of the anomers, and then stored at 4 °C. Stock solutions of 10 mM HP and 100 mM AA were prepared in distilled water and 100 mM HCl, respectively.

2.2. Instrumentation and Software

Calibrations for HP, AA and glucose were performed in a standard three-electrode cell containing 20 mL PBS at room temperature, a saturated calomel reference electrode (SCE), a stainless steel auxiliary electrode and either bare or modified platinum-iridium (90:10) working electrodes. Constant potential amperometry was performed at an applied potential of +0.7 V versus SCE, using Chart (v 5.2) software (AD Instruments Ltd., Oxford, UK) and a low-noise potentiostat (Biostat IV, ACM Instruments, Cumbria, UK). The working electrodes were allowed to settle in quiescent PBS to give a steady background current before the addition of small known aliquots of the analyte of interest.

2.3. Working Electrode Preparation

Cylinder electrode preparation has been described in detail recently [44]. Briefly, 125 μm diameter Teflon-coated Pt-Ir wire (90:10, Advent Research Materials Ltd., Eynsham, England) was stripped of 1 mm Teflon to expose the bare metal, which displays many of the electrochemical properties of pure Pt [44]. Electropolymerization was carried out in oPD solutions (of varied monomer concentration, background electrolyte and enzyme concentration) at +0.7 V versus SCE for 15 minutes for these Pt_C electrodes [39,41]. Three main enzyme immobilization protocols were used in this work. In the first, the enzyme was immobilized by adsorption and dip-evaporation before PoPD deposition [23]. Each electrode was dipped in a 200 U·mL^{-1} solution of GOx for 5 minutes, allowed to dry for 5 min, and then dipped quickly into the GOx

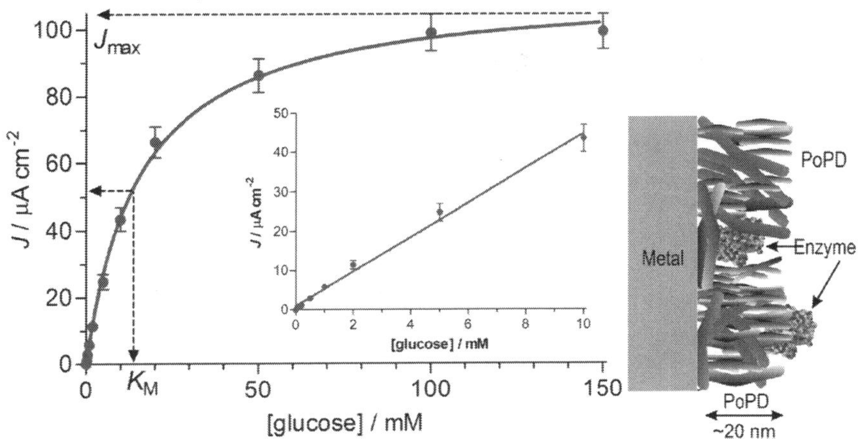

Figure 1. Sample steady-state calibration data and nonlinear regression analysis for the biosensor design, Pt_C/PoPD-GOx [Equation (4), R^2 = 0.998, n = 8; **left**], illustrating the graphical significance of the Michaelis-Menten constants, J_{max} and K_M. The linear region slope (LRS) was obtained using linear regression up to 10 mM glucose (R^2 = 0.996, n = 8; **left inset**), and represents the most suitable measure of analytical sensitivity of each biosensor design to enzyme substrate (see Table 1). Schematic representation of the PEC configuration for the same Pt_C/PoPD-GOx design (**right**), illustrating trapped GOx (~8 nm diameter) in the PoPD layer deposited by the precipitation of insoluble chains formed during the electropolymerization of monomer solution containing the enzyme.

Table 1. Mean ± SEM (n = number of electrodes) for the two apparent Michaelis-Menten parameters J_{max} and K_M determined using nonlinear regression and Equation (4) for glucose calibrations (see Figure 1), and the corresponding linear region slope (LRS) values. PoPD-based biosensors were electrosynthesized in PBS containing 300 mM oPD and no enzyme, except for the Pt_C/PoPD-GOx design which included 5 mg mL^{-1} GOx. The PoPD-free design is included for comparison.

Design	*n*-value	J_{max} (µA cm^{-2})	K_M (mM)	LRS (µA cm^{-2} mM^{-1})	BE% (%)
Pt_C/GOx-GA	4	66 ± 5	13 ± 1	3.6 ± 0.2	1.4 ± 0.1
Pt_C/GOx-GA/PoPD	4	33 ± 3	32 ± 3	1.2 ± 0.1	0.47 ± 0.05
Pt_C/GOx/PoPD	7	10 ± 1	21 ± 2	0.37 ± 0.03	0.17 ± 0.01
Pt_C/PoPD/GOx-GA	4	72 ± 3	26 ± 2	2.7 ± 0.1	0.66 ± 0.02
Pt_C/GOx/PoPD/GOx-GA	4	50 ± 5	26 ± 3	1.7 ± 0.4	0.55 ± 0.07
Pt_C/PoPD-GOx	8	111 ± 6	16 ± 2	5.0 ± 0.4	2.0 ± 0.2

solution four more times with 5 minutes drying between each dip, followed by electropolymerization. This protocol was previously found to optimize enzyme loading for biosensors of the type Pt_C/EOx/PoPD [24]. The second design immobilized the enzyme by adsorption and dip-evaporation after PoPD deposition followed by exposure to glutaraldehyde (GA) vapour for 15 min to crosslink the enzyme [23], and are termed Pt_C/PoPD/GOx-GA. The third method used co-immobilization, whereby either 1 mg·mL^{-1} (~650 U·mL^{-1}; ~5 μM) or 5 mg·mL^{-1} GOx was dissolved in oPD, and electropolymerized at +0.7 V vs. SCE for 15 min [30,31] to give Pt_C/PoPD-GOx; see Figure 1.

2.4. Enzyme Kinetic Parameters

First generation biosensors of the general design Pt_C/PoPD~EOx (i.e., various configurations of enzyme deposited before, over, or simultaneously with, the PoPD layer) display Michaelis-Menten kinetics, as discussed recently [11]. A previous study has shown that substrate diffusion is not limiting for non-conducting PoPD layers incorporating enzyme [32], due to their ultrathin nature (10–30 nm [31,45,46]). Therefore, the basic Michaelis-Menten enzyme parameters used here provide more readily accessible insights into factors affecting the responsiveness of biosensors fabricated from this polymer, and avoids the use of more complex analyses such as those involving the Thiele modulus [47].

Although a ping-pong mechanism describes the enzyme interaction with the substrate and co-substrate [Equations (1) and (2)], the oxygen effect was not included in the analysis here for simplicity. This is justified on two grounds: the concentration of oxygen was fixed in the present experiments (air saturation); and oxygen effects on biosensors of these designs are minimal for the range of substrate and oxygen levels encountered during neurochemical monitoring in vivo for both glucose [23,48] and glutamate [27,28]. Thus, the one-substrate form of the Michaelis-Menten equation contains the parameters used here to compare biosensor performance [Equation (4)], where J_S is the current-density normalized, background subtracted, biosensor response to a specified concentration of enzyme substrate ([S]):

$$J_S = \frac{J_{max}}{1 + K_M/[S]} \tag{4}$$

J_{max} is the maximum, or plateau, current density response, obtained when all enzyme sites are saturated with substrate (see Figure 1). Different values of J_{max}, determined under the same conditions, reflect differences in the amount of active (not total) enzyme on the surface, provided k_{cat} and the sensitivity of the electrode to H_2O_2 [Equation (3)] does not vary much [11]. K_M is the apparent Michaelis constant, and phenomenologically defines the concentration of substrate that gives half the J_{max} response. Changes in K_M are sensitive to variations in enzyme-substrate access/binding, and have been interpreted in terms of barriers to enzyme-substrate access [26,49], as it is in the present study. K_M is also useful for defining the range of the linear response to S (up to ~½K_M), as well as determining the linear region slope (LRS), i.e., LRS ≈ J_{max}/K_M for a true hyperbolic response [11,27]. A plot of J_S versus [S] up to enzyme saturation therefore provides the basic kinetic parameters J_{max} and K_M, as well as the nonlinear coefficient of determination, R^2; a similar plot up to ~½K_M provides the analytically key substrate sensitivity parameter, LRS, and the linear coefficient of determination, R^2 (see Figure 1). To account for any variations in LRS caused by differences in polymer-enzyme composite (PEC) biosensor sensitivity to HP, the parameter BE% was defined as the HP-normalized LRS [Equation (5)], which can be considered to reflect the efficiency of the biosensor in converting substrate to HP [11]. This parameter also allows the efficiency of the enzyme layer to be compared across diverse biosensor designs:

$$BE\% = \frac{\text{LRS at Pt/PEC}}{\text{slope(HP) at Pt/PEC}} \times 100\% \tag{5}$$

2.5. Permeability and Permselectivity Parameters

From the calibration plots for HP, AA and glucose, a number of parameters were calculated to quantify the performance of the different designs. The apparent analyte permeabilities to HP and AA were calculated from Equations (6) and (7) [11], which is similar to other studies [50], and discussed in detail recently [41]:

$$P(\text{HP})\% = \frac{\text{slope(HP) at Pt/PEC}}{\text{slope(HP) at bare Pt}} \times 100\% \tag{6}$$

The slopes (μA cm^{-2} mM^{-1}) of the linear responses for HP on bare metal and on the PoPD-modified electrodes were obtained from linear regression analysis of the respective calibration plots of the steady-state HP responses versus HP concentration up to 0.1 mM, as were the AA slopes up to 1 mM for the bare electrodes. In contrast, the AA response was nonlinear and self limiting (see Figure 2), as observed previously for similar PoPD-based designs [30,41], and linked with "self-blocking" by AA-related species trapped in the polymer matrix [45]. Hence, the current density at 1 mM AA [J_{AA}(1 mM)] was used as a measure of the AA response (see Figure 2). This AA concentration is relevant to neurochemical applications because baseline brain AA levels are ~0.5 mM [51], reaching millimolar levels during periods of behavioral stimulation [52–54]. The ideal values of P(HP)% and P(AA)% for biosensor applications are therefore

100% and 0%, respectively. In addition, the concentration of other electroactive interference compounds in the brain, such as dopamine and its metabolites [55], are orders of magnitudes smaller than that of AA, ensuring that their contributions to PPD-based biosensor responses are insignificant [24,25,56–58]:

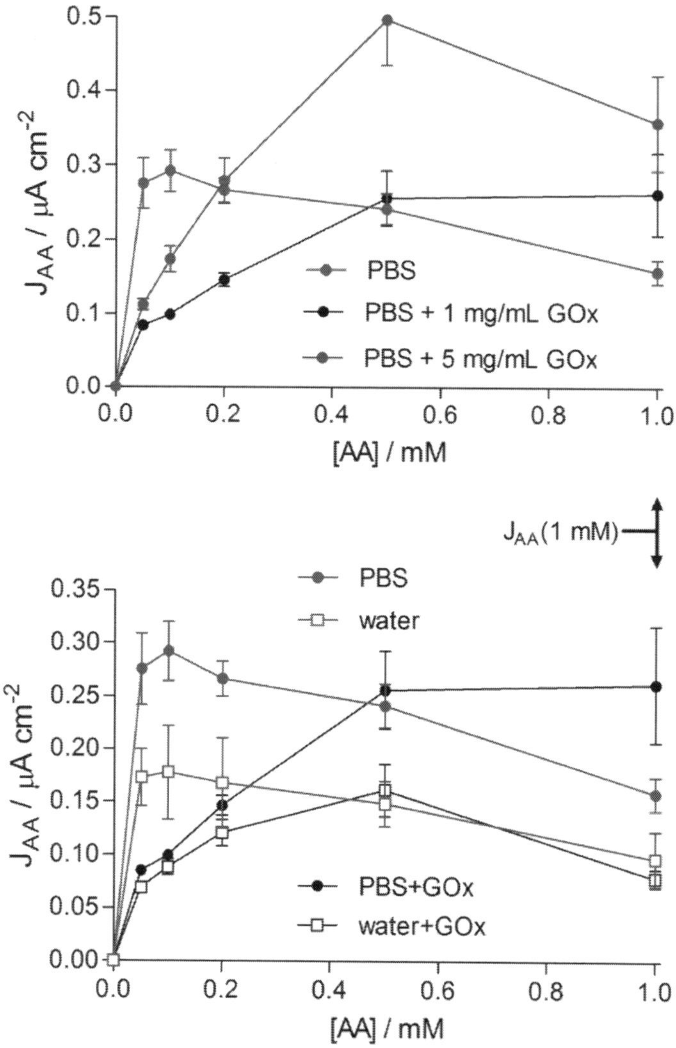

Figure 2. Averaged steady-state AA calibrations for Pt_C/PoPD and Pt_C/PoPD-GOx electrosynthesized in 300 mM oPD solution made with either PBS*, PBS + GOx (1 mg mL^{-1}, n = 4), PBS+GOx (5 mg mL^{-1}, n = 8), distilled water*, or water+GOx (1 mg mL^{-1}, n = 6). The concentration of GOx for the bottom graph was 1 mg mL^{-1}. *The GOx-free data were taken from the literature [38] for comparison.

$$P(\mathrm{AA})\% = \frac{J_{\mathrm{AA}}(1\,\mathrm{mM})\ \mathrm{at\ Pt/PEC}}{J_{\mathrm{AA}}(1\,\mathrm{mM})\ \mathrm{at\ bare\ Pt}} \times 100\% \tag{7}$$

A polymer selectivity parameter, S% (Equation 8), has been defined as the percentage interference by AA in HP detection for equimolar concentrations [41,44], with an optimum value of 0%:

$$S\% = \frac{J_{\mathrm{AA}}(1\,\mathrm{mM})\ \mathrm{at\ Pt/PEC}}{J_{\mathrm{HP}}(1\,\mathrm{mM})\ \mathrm{at\ Pt/PEC}} \times 100\% \tag{8}$$

Although S% has been a useful parameter for gaining insights into the performance of the enzyme-free electrosynthesized polymer [11], it is not a sufficient index of the selectivity of PEC-based biosensors. Hence $S_S\%$, the equimolar enzyme substrate selectivity parameter [Equation (9), e.g., $S_G\%$ for glucose], has also been described [11], which is similar to, but more straightforward than, non-equimolar equivalents [50,59]. The ideal value of $S_S\%$ is zero, and compares the capacity of the PEC layer to generate current from the enzyme reactions, J_S [see Equations (1–3)], to the interference response produced by AA, J_{AA}, for 1 mM of each analyte, a concentration which is close to brain extracellular fluid values in vivo for both compounds [51,60]:

$$S_S\% = \frac{J_{\mathrm{AA}}(1\,\mathrm{mM})\ \mathrm{at\ Pt/PEC}}{J_S(1\,\mathrm{mM})\ \mathrm{at\ Pt/PEC}} \times 100\% \tag{9}$$

The permeability and selectivity parameters, as well as BE% [Equations (5–9)], reflect intrinsic properties of PoPD which are normalized with respect to actual electrochemical surface area, rather than geometrically calculated area. All parameters were determined for individual electrodes and then averaged over populations of sensors for each design. Results are reported are mean ± standard error (SEM), with n = number of electrodes. Currents are presented as current density, calculated using the geometric area of these smooth wire electrodes. Linear and nonlinear regression analyses were performed using GraphPad Prism (version 5.02, San Diego, CA, USA). The statistical significance of variations between parameters for the different designs was calculated using Student's two-tailed unpaired t-tests (Prism 5.02), with values of $p < 0.05$ considered to indicate statistical significance of the difference.

3. RESULTS AND DISCUSSION

A large body of work on GOx-based biosensors incorporating an electrosynthesized PPD permselective layer for glucose detection has been published over the past two decades, and a limited selection is cited here [10,11,18,20,26,31,32,38,39,46,61–63]. However, a broad range of variables are involved in the fabrication and characterization of these biosensors, including: which of the three phenylenediamine monomers is used; the concentration of monomer; the background electrolyte and pH of the electropolymerization medium; the choice of cyclic voltammetry or fixed applied potential (and the value of applied potential) in the PPD electrosynthesis step; the size and shape of the electrode substrate; the mode of enzyme immobilization and its concentration; the variety and concentration range of the interference species studied; flow versus quiescent calibration systems; surface imaging and spectrochemical characterization; etc. Therefore, further optimization of this system is possible, and would be useful both in terms of understanding the nature of the polymer-enzyme composite (PEC) layer (see Figure 1) and improving the performance of the biosensor device.

In our laboratory, significant enhancement of the glucose sensitivity for a PoPD-based biosensor has been described, by using a Pt disk geometry (Pt_D) to increase GOx loading [23]. However, more recently, precise permeability data revealed a novel edge effect which compromised the ability of Pt_D/PoPD devices to block interference [41]. Taken together, these reports highlight the need to integrate enzyme kinetic analysis and detailed polymer permselectivity in the characterization of specific biosensors [64]. The enzyme substrate selectivity parameter, S_S% [Equation (9)], is a key measure of the balance between high enzyme sensitivity and low interference responses needed for practical biosensors. A number of literature studies have used a similar, non-equimolar, version of this parameter to good effect [39,50,59,65]. Here we apply an extensive range of enzyme kinetic and polymer permeability/permselectivity parameters, including S_S%, to characterize and further optimize the properties of a low edge-density Pt_C/PoPD-based biosensor electrosynthesized under different conditions, including the novel environment of no added background electrolyte [42].

3.1. Michaelis-Menten Characteristics of the Basic Designs

Three main methods of GOx immobilization were examined here: dip-evaporation before polymer electrosynthesis (Pt_C/GOx/PoPD), dip-evaporation over the polymer using glutaraldehyde (GA) as a crosslinker (Pt_C/PoPD/GOx-GA), and co-polymerization from the monomer solution (Pt_C/PoPD-GOx). Glucose calibrations for biosensors of all designs followed Michaelis-Menten hyperbolic behavior [Equation (4), see Figure 1]; the corresponding J_{max}, K_M and

LRS values for biosensors fabricated using our standard background electrolyte and monomer concentration (PBS containing 300 mM oPD [24,30]) are given in Table 1, as well as parameters for the PoPD-free design, Pt_C/GOx-GA, for comparison.

The PoPD-free design (Pt_C/GOx-GA) displayed a moderate J_{max} value indicative of good active enzyme loading. The corresponding K_M value was the lowest of these six designs, indicating that the enzyme was readily accessible to the substrate, although values lower than 5 mM have been reported previously [66,67]. Addition of the PoPD layer after enzyme deposition (GOx/PoPD) decreased the J_{max} and increased the K_M, the latter indicating that PoPD hindered access of the substrate to the surface-bound enzyme, as observed previously for a PoPD layer containing the protein, bovine serum albumin, BSA [23]. The decrease in J_{max} following deposition on the PoPD could be due to either covering of the GOx by the polymer, or displacement of the enzyme off the surface during the PoPD deposition. The finding that the J_{max} for Pt_C/GOx-GA/PoPD was three-fold greater than that for Pt_C/GOx/PoPD (see Table 1) suggests that, in the absence of crosslinking with GA, much of the enzyme is removed from the surface by electro-deposition of the polymer. This interpretation is consistent with the ultrathin nature of electrosynthesized PoPD [31,45,46] which is considered not to overwhelm immobilized enzyme [31,68]; see Figure 1.

Immobilization of GOx over the PoPD layer (PoPD/GOx) showed a 7-fold increase in active enzyme loading compared with the GOx/PoPD configuration, and was indistinguishable from the PoPD-free configuration ($p > 0.34$). Surprisingly, the K_M value was similar for PoPD/GOx and GOx/PoPD, suggesting that enzyme-substrate binding was compromised by the presence of PoPD (see Section 3.2). In an attempt to increase GOx loading further, GOx was dip-evaporated both before and after polymer fabrication (GOx/PoPD/GOx), with no benefit obtained (Table 1). Neither was any significant enhancement in J_{max} observed when a 2,000 U mL^{-1} GOx solution was used in the dip/evaporation procedure (data not shown).

In the final basic design, GOx (5 mg mL^{-1}) was dissolved in the monomer solution for co-immobilization during electropolymerization, as described previously for other conditions and electrode geometries [30–32,50]. This PoPD-GOx configuration showed the highest active enzyme loading, nearly twice the value of the PoPD-free design ($p < 0.001$). The mean K_M value for Pt_C/PoPD-GOx was not statistically different from the PoPD-free design ($p > 0.33$), indicating similar substrate-enzyme access. In addition, K_M for the PoPD-GOx configuration was significantly lower than for PoPD/GOx ($p < 0.02$). One speculation is that the different structure of PoPD deposited in the presence of solution GOx, which has been observed in scanning electron microscopy studies [30], is less obstructive to substrate binding, a notion supported by AA permeability data below (see Section 3.2). Irrespective of the mechanism, however, clearly the co-immobilization of GOx from the monomer solution is superior in terms of active enzyme loading and affinity (see Table 1) compared with these, and other BSA-containing, Pt_C/PoPD~GOx biosensor designs [23]. This detailed comparison supports the protocols advanced previously for the co-immobilization of enzyme with PoPD for biosensor fabrication [30–32,48,50,68].

The linear region slope (LRS) of the glucose calibration, a parameter determined by both the J_{max} and K_M values [11], is a better index of the functional sensitivity of the different designs. As expected, the deposition of GOx before the polymer led to the lowest LRS sensitivity, ~8-fold lower than incorporation of enzyme over the polymer. This trend is in line with that reported recently, where BSA was incorporated in the PoPD matrix [23]. Co-polymerization of GOx displayed the highest LRS sensitivity, twice as good as the next ranking PoPD-based biosensor, Pt_C/PoPD/GOx-GA ($p < 0.003$). LRS values are influenced by two main factors: the ability of the enzyme layer to convert substrate to HP (Equations (1)–(2)), and the sensitivity of the electrode to HP [Equation (3)]. This latter can be determined as the biosensor HP calibration slope, and normalization of LRS with respect to this HP slope provides an index of the efficiency of the biosensor to convert substrate to HP [BE%, Equation (5)]. As well as being of intrinsic interest, this parameter becomes of practical importance when biosensors are used in environments where HP is produced by other components in the medium, such as mitochondria in brain tissue [69]. The maximum value of BE% for the designs shown in Table 1 was ~2% for the Pt_C/GOx-GA and Pt_C/PoPD-GOx configurations. This low value contrasts with estimates of ~50% for PoPD-based glutamate biosensors, mainly due to the much higher affinity of glutamate oxidase for its substrate compared with the GOx-glucose system [70].

Response times were recorded in constantly stirred solution, using the PowerLab module operating at a data acquisition rate of > 100 Hz. A $t_{90\%}$ parameter was defined as the time taken for the analyte response to reach 90% of its maximum value from the start of the current upswing, and is similar to definitions used previously [71–73]. The co-immobilized PoPD-GOx design was used in this study because of its high enzyme loading and LRS sensitivity (Table 1). The response time for glucose was fast ($t_{90\%} = 1.7 \pm 0.1$ s, n = 6), with the corresponding response time for injections of HP aliquots ($t_{90\%} = 1.3 \pm 0.1$ s) indicating that only ~0.4 s of the glucose response was attributable to the enzyme reactions [Equations (1–2)]. This compares favorably with a glutamate Pt_C-based biosensor where glutamate oxidase was in the PoPD layer, and the glutamate component of the response time was ~0.6 s [27]. These results are also consistent with previous time-response studies of the PoPD-GOx system [31], and with the ultrathin nature of PoPD, electrosynthesized under non highly-acidic conditions, allowing fast interaction of the enzyme with its substrate (see Figure 1). It appears, therefore, that K_M is a much more sensitive index of hindrance in enzyme-substrate interactions for these glucose biosensors (Table 1), as in the case of PoPD-based glutamate biosensors, where little difference in response time was observed across diverse PEC configurations with largely different K_M values [27].

3.2. Permeability Characteristics of the Basic Designs

Good permeability of the PEC membrane to HP is important for practical first-generation biosensor designs. The apparent HP permeability, P(HP)% defined by Equation (6), was similar for all designs (Table 2), with an average value of 106 ± 8% (n = 41) which was not significantly different from the ideal value of 100% ($p > 0.42$). Therefore differences in the permselectivity parameter, S% [Equation (8)], across the designs should be influenced mainly by polymer interference-rejection properties. Similarly, differences in the biosensor selectivity parameter, S_S% [Equation (9)], should be due to a combination of polymer interference rejection and the ability of the PEC layer to generate HP. The finding that some P(HP)% values were greater than 100% is unexpected, but has been observed before for PoPD layers containing a variety of macromolecular modifiers [44]. These supra-optimal values may be due minor disproportionation of HP on metals, and its possible inhibition by polymer coatings [74].

Table 2. Mean values ± SEM (n = number of electrodes) for the two apparent permeabilities, P(HP)% and P(AA)% determined using Equations (6) and (7), respectively, and for the two selectivity parameters, S% and S_G%, defined by Equations (8) and (9), respectively, for both PEC-coated and enzyme-free PoPD-modified Pt_C electrodes. The electropolymerization solution contained 300 mM oPD in PBS, and 5 mg mL^{-1} GOx for the co-immobilization design (Pt_C/PoPD-GOx).

Design	*n*-value	J_{max} (µA cm^{-2})	K_M (mM)	LRS (µA cm^{-2} mM^{-1})	BE% (%)
Pt_C/GOx-GA	4	66 ± 5	13 ± 1	3.6 ± 0.2	1.4 ± 0.1
Pt_C/GOx-GA/PoPD	4	33 ± 3	32 ± 3	1.2 ± 0.1	0.47 ± 0.05
Pt_C/GOx/PoPD	7	10 ± 1	21 ± 2	0.37 ± 0.03	0.17 ± 0.01
Pt_C/PoPD/GOx-GA	4	72 ± 3	26 ± 2	2.7 ± 0.1	0.66 ± 0.02
Pt_C/GOx/PoPD/GOx-GA	4	50 ± 5	26 ± 3	1.7 ± 0.4	0.55 ± 0.07
Pt_C/PoPD-GOx	8	111 ± 6	16 ± 2	5.0 ± 0.4	2.0 ± 0.2

The apparent AA permeability, P(AA)%, was calculated using Equation (7); all PoPD-modified designs blocked the 1 mM AA flux by ≥ 99% compared with the bare metal, with P(AA)% ≤ 1% (Table 2). The best blocking characteristics were displayed by Pt_C/PoPD. i.e., by the pure PoPD, with P(AA)% = 0.11 ± 0.02%, a value similar to that reported for PoPD deposited under the same conditions on pure, low edge-density, Pt microfiber electrodes [41]. The finding that P(AA)% for Pt_C/PoPD/GOx (0.34 ± 0.05%, $p < 0.001$) was significantly greater than for the native polymer (Pt_C/PoPD) suggests that the enzyme does not simply sit on top of the PoPD, but imbeds in the polymer, opening its structure and undermining its interference blocking to a small degree. Surprisingly, the GOx deposited before the PoPD had a similar effect on P(AA)% compared with the PoPD/GOx configuration. This view is consistent with the K_M data in Table 1, which shows that the barrier to GOx-glucose interactions was significantly greater for Pt_C/PoPD/GOx compared with Pt_C/GOx ($p < 0.002$), but not with Pt_C/GOx/PoPD ($p > 0.13$). It is interesting to note that P(AA)% for the GOx/PoPD and PoPD/GOx configurations was additive when compared with Pt_C/GOx/PoPD/GOx, the worst AA-rejecting configuration (1.0 ± 0.1%). This indicates that successive fabrication steps (in this case, GOx deposited both before and after the PoPD) perturb the polymer, compounding the slight undermining of its interference blocking properties. Finally, P(AA)% for the co-immobilized enzyme (GOx-PoPD, 0.24 ± 0.04%), which had the best glucose LRS value (Table 1), was second only to pure PoPD in terms of AA rejection (99.76%) for these basic designs.

To investigate whether the presence of GOx on the metal surface (GOx/PoPD design) or in the monomer solution (PoPD-GOx design) affected the rate of deposition of the PoPD, the collapse of electropolymerization current associated with the self-sealing nature of this polymer as it deposits on the electrode surface was analysed. The anodic electropolymerization current fell off very rapidly following the initial surge induced by the application of 0.7 V versus SCE, leading to a ~99% loss of initial current by ~10 s for electropolymerizations carried out in PBS containing 300 mM oPD. A 2-phase exponential decay model gave a significantly better nonlinear regression fit compared with a 1-phase analysis, as observed [22] and discussed [75] recently. The half-life values for the associated two time domains, $t_{½}$(fast) and $t_{½}$(slow), are a measure of the rate at which the blocking layer of PoPD builds up on the metal, and so might be expected to be influenced by the presence of protein macromolecules near the electrode surface. The reference values determined in the absence of GOx (i.e., for Pt_C/PoPD) were 60 ± 10 ms and 0.40 ± 0.02 s ($n = 8$), respectively. The presence of GOx on the surface prior to electro-deposition (Pt_C/GOx/PoPD) did not significantly affect the rate of electropolymerization current decay: 50 ± 10 ms and 0.48 ± 0.06 s ($n = 11$, $p > 0.28$). In contrast, the present of GOx (5 mg mL^{-1}) in the monomer solution did significantly slow the current collapse in both time domains: 140 ± 10 ms and 0.93 ± 0.06 s ($n = 8$, $p < 0.001$). Work is currently underway to understand more fully the significance of these fast and slower components of PoPD electrosynthesis [75], but here, as in recent findings [42], there does not appear to be any correlation between the rate of the electropolymerization current collapse and the apparent permeability of AA in the PoPD deposited (see Table 2 and Section 3.3).

As expected from the corresponding definitions and the relatively constant value of P(HP)% across the biosensor designs studied here (Table 2), the trend in the permselectivity parameter (S%), calculated using Equation (8), paralleled that of P(AA)% (see Table 2). It is the equimolar substrate selectivity (S_G%), defined by Equation (9), which most clearly reveals the superiority of the co-immobilized configuration among these basic designs (Table 2). The mean S_G% value for Pt_C/PoPD-GOx (7 ± 1%, n = 8) was between 4 and 50 times smaller (better) than for the other designs. Thus, in media containing equal concentrations of glucose and AA at the ~1-mM level, as is the case in brain extracellular fluid [51,60], the baseline response of the Pt_C/PoPD-GOx biosensor would have a ~7% interference contribution from AA. However, because changes in the biosensor signal are far more important than their absolute output, interference in monitoring glucose changes would be considerably less than this value due to the self-limiting shape of the AA response (see Figure 2).

3.3. Fine-Tuning the Conditions for Co-Immobilization of PoPD and GOx

Biosensors fabricated using the co-immobilization conditions described above show more than adequate substrate selectivity for glucose monitoring in most biological fluids. That said, it is always desirable to reduce biosensor interference as far as possible, especially in the design of biosensors of substrates which exist at much lower levels, such as when monitoring the key neurotransmitters, glutamate [76,77] and acetylcholine [78,79]. Electropolymerization conditions were therefore modified in attempts to lower further S_G% for Pt_C/PoPD-GOx devices compared with the standard conditions of 300 mM oPD in PBS containing 5 mg mL^{-1} GOx.

3.3.1. Monomer Concentration

Previous studies have shown that there is little difference between the interference rejection properties of PoPD formed at widely different concentrations [24], and effective permselective PPD layers are often generated from solutions with monomer concentrations as low as 5 mM [19,31,46,46,80,81] and 3 mM [26,46,58]. However, for the detailed analysis and comparisons of the present study, 10 mM and 100 mM oPD were tested for the most selective basic design (Table 2), i.e., co-immobilization of PoPD with 5 mg mL^{-1} GOx. In line with these cited reports, there was little difference between P(AA)% for all three oPD concentrations. Thus, although the highest concentration tested (300 mM, which is close to saturation) displayed the lowest (best) AA permeability (0.24 ± 0.04%, n = 8), there was no significant difference between this and P(AA)% determined for Pt_C/PoPD-GOx biosensors electrosynthesized in 10 mM oPD (0.33 ± 0.05%, n = 4, $p > 0.2$).

However, large differences in the glucose Michaelis-Menten parameters were observed as a function of monomer concentration. Active enzyme loading was highest for 300 mM oPD (J_{max} = 111 ± 6 μA cm^{-2}, n = 8; see Figure 1) and lowest for 10 mM (3.4 ± 0.5 μA cm^{-2}, n = 4, $p < 0.001$). The enzyme affinity was also greatest for 300 mM monomer (K_M = 16 ± 2 mM, n = 8) compared with 10 mM oPD (K_M = 36 ± 3 mM, n = 4, $p < 0.001$). These data are consistent with the notion that such a low population density of GOx molecules in the PEC layer would be more hindered by the polymer, as observed previously in a detailed analysis of the correlation between GOx loading and K_M for Pt_C-based glucose biosensors [23]. These two factors (enzyme loading and affinity) combined to produced a 50-fold decrease in the LRS for biosensors fabricated in 10 mM oPD (LRS = 0.09 ± 0.02 μA cm^{-2} mM^{-1}, n = 4) compared with 300 mM monomer (5.0 ± 0.4 μA cm^{-2} mM^{-1}, n = 8, $p < 0.001$), which is the opposite trend reported for PoPD-based glucose biosensors made in different oPD concentrations, using an FIA system [50]. The oPD concentration was therefore maintained at 300 mM throughout the remainder of this investigation.

3.3.2. Enzyme Concentration

The polymerization conditions for the basic co-immobilization design (Figure 1, and Tables 1 and 2) involved 5 mg mL^{-1} (~3 kU mL^{-1}) GOx in the monomer solution because this enzyme concentration had been adopted in previous studies to optimise selectivity for disk-based biosensors fabricated in PBS [30]. Given our greater understanding now of the different factors affecting the performance characteristics of disk versus cylinder biosensors [23,27,28,41,70], and the more common use of lower enzyme activity solutions for co-immobilization of GOx [31,50,82,83], Pt_C/PoPD-GOx biosensors made from 1 mg mL^{-1} GOx (~650 U mL^{-1}; ~5 μM) in PBS containing 300 mM oPD were characterized (see Table 3). There was only a small, but statistically significant, decrease in glucose calibration mean J_{max} values for biosensors fabricated in 1 mg mL^{-1} versus 5 mg mL^{-1} GOx solutions ($p < 0.04$), with no significant difference between the K_M values ($p > 0.4$). Not surprisingly, therefore, both the mean LRS and BE% values were indistinguishable for the two populations of biosensors ($p > 0.9$). Moreover, the finding that the 1-mM AA rejection parameter P(AA)% was not significantly different ($p > 0.3$; see Figure 2) meant that S% and S_G% were also indistinguishable for the two groups (Table 3).

Table 3. Mean values ± SEM for the apparent Michaelis-Menten parameters J_{max} and K_M determined using nonlinear regression [Equation (4)] for glucose calibrations and the linear region slope (LRS) for Pt_C/PoPD-GOx biosensors electrosynthesized in different media. The apparent AA permeability P(AA)% determined using Equation (7) and the two selectivity parameters, S% and S_G%, determined using Equations (8) and (9), respectively, for these biosensors fabricated in 300 mM oPD and 1 mg mL^{-1} GOx (unless stated otherwise) dissolved in different added background electrolytes: PBS (5 mg mL^{-1} GOx, n = 8), PBS (n = 4), KCl (n = 4) and no added background electrolyte (distilled water, n = 6).

Background electrolyte:	**PBS**	**PBS**	**KCl**	**none**
[GOx] (mg mL^{-1}):	5.0	1.0	1.0	1.0
J_{max} (μA cm^{-2})	111 ± 6	85 ± 7	87 ± 3	96 ± 5
K_M (mM)	16 ± 2	13 ± 2	27 ± 1	10 ± 1
LRS (μA cm^{-2} mM^{-1})	5.0 ± 0.4	5.0 ± 1.0	3.1 ± 0.2	6.5 ± 0.6
BE%	2.0 ± 0.2	2.0 ± 0.4	1.1 ± 0.1	2.4 ± 0.2
P(AA)%	0.24 ± 0.04	0.18 ± 0.04	0.11 ± 0.01	0.11 ± 0.02
S%	0.14 ± 0.02	0.09 ± 0.02	0.06 ± 0.01	0.07 ± 0.01
S_G%	7 ± 1	7 ± 2	6 ± 1	2 ± 1

There were, however, subtle effects of different enzyme concentrations in the polymerization solution on the subsequent AA calibration responses (see Figure 2, top). When no protein was present, the maximum steady-state AA current was observed at low AA levels (~0.1 mM), and the response diminished gradually and steadily thereafter. When 1 mg mL^{-1} GOx was included in the polymerization medium (i.e., for Pt_C/PoPD-GOx_1), the AA calibration was more hyperbolic with significantly smaller AA currents at lower AA concentrations. The higher concentration of enzyme (5 mg mL^{-1}) in the monomer solution (Pt_C/PoPD-GOx_5), however, led to greater responses for all AA concentrations compared with the 1 mg mL^{-1} level (Figure 2, top). Therefore, because the additional enzyme activity in the polymerization medium did not increase substrate sensitivity for the resulting biosensors, and because the shape of the AA calibrations recorded with Pt_C/PoPD-GOx_1 were more benign for applications involving AA-containing biological media, the 1 mg mL^{-1} GOx concentration in 300 mM oPD was used throughout the remainder of this work.

3.3.3. Electropolymerization Background Electrolyte

A recent study showed that omission of added background electrolyte from the oPD (weak electrolyte) solution slowed down the electropolymerization current collapse by two orders of magnitude, but surprisingly augmented the blocking ability of the enzyme-free Pt_C/PoPD formed [42]. Biosensors were therefore fabricated here in solutions of 1 mg mL^{-1} GOx dissolved in distilled water containing 300 mM oPD, and characterized in terms of Michaelis-Menten, permeability and selectivity parameters (see Table 3). The collapse in the electropolymerization current was slow in the absence of added background electrolyte for Pt_C/PoPD(H_2O), with a 1-phase exponential decay half-life value of 228 ± 20 s, n = 9. Incorporation of 1 mg mL^{-1} GOx in the 300 mM oPD distilled water solution marginally accelerated the formation of the self-sealing polymer layer (half life value of 159 ± 12 s, n = 3, $p < 0.09$), reflecting the increase in solution conductivity caused by the presence of the protein polyelectrolyte.

There was no significant difference between any of the enzyme parameters, including BE%, for biosensors generated in PBS compared with no added background electrolyte (Table 3). In line with previous enzyme-free studies [42], the mean P(AA)% for Pt_C/PoPD-GOx(H_2O) was ~40% less than that determined for Pt_C/PoPD-GOx(PBS). There was also no significant difference ($p > 0.8$) between P(AA)% determined for Pt_C/PoPD(H_2O) versus Pt_C/PoPD-GOx_1(H_2O), indicating that the ~5 μM level of polyelectrolyte enzyme was low enough not to affect the structure of the PoPD in such a way as to influence this parameter. In addition to this improvement in AA blocking at 1 mM levels, there was a decrease in the AA response across the entire concentration range of the AA calibration (see Figure 2, bottom). In view of these observations and the finding that AA responses at low AA levels were significantly decreased by this concentration of GOx in the polymerization solution for both PBS and no background electrolyte conditions (Figure 2), there is clearly much remaining to be understood about the structure of surface PoPD electro-deposited in non-acidic media [30,84–86], the influence of trapped enzymes, and details of the interactions of the PEC layer with AA during calibrations. At the analytically useful phenomenological level, however, there was a ~3-fold improvement in S_G% for biosensors fabricated in the absence of background electrolyte (Table 3).

In view of these, and published [42,87], findings that the composition of the background electrolyte affects the permeability of even the non-conducting, non-ionic form of this polymer electrogenerated at non-acidic pH [88], the effects of different alkali metal salts in the monomer solution were investigated in the first instance for the enzyme-free Pt_C/PoPD devices. Figure 3 (top) shows the trend in P(AA)% for these three electrolytes at 150 mM levels in the polymerization solution. There was a consistent decrease (improvement) in AA permeability with decreasing hydrated cation radius, with the optimum value

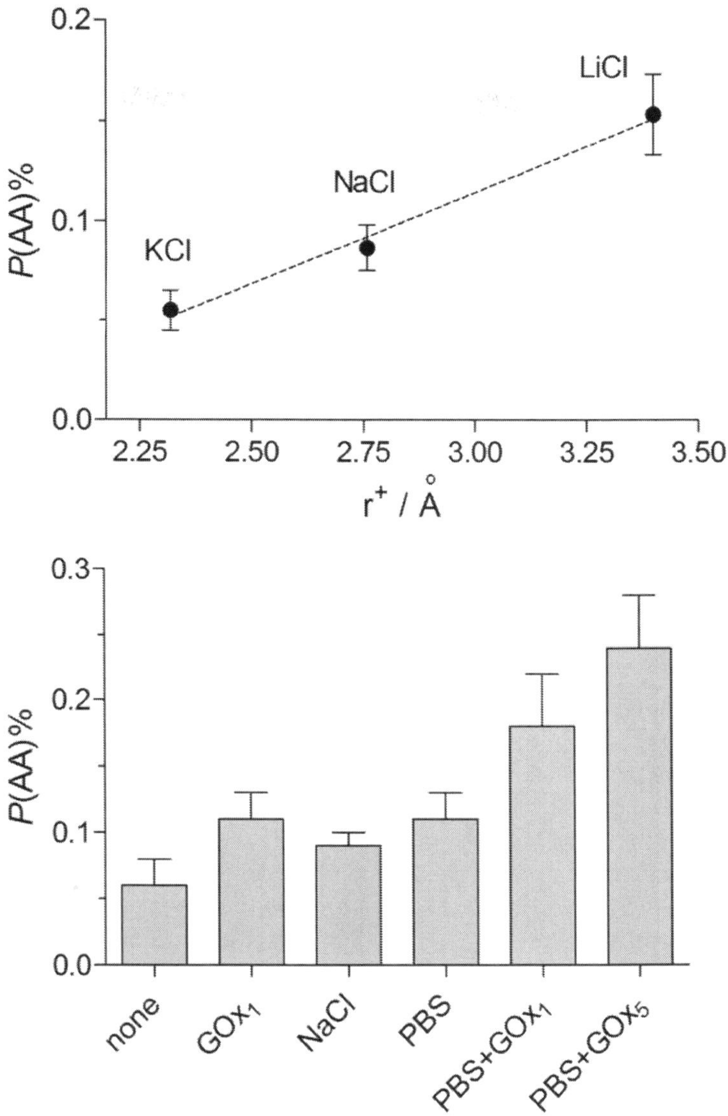

Figure 3. Effect of different background electrolytes and GOx concentrations in the monomer solution on the subsequent P(AA)% values determined for Ptc/PoPD electrodes electrosynthesized from 300 mM oPD. **Top:** 150 mM of either KCl (n = 12), NaCl (n = 8) or LiCl (n = 4) plotted against the hydrodynamic radius of the cations. **Bottom** (left to right): no added background electrolyte (i.e., distilled water, n = 28); distilled water containing 1 mg mL^{-1} GOx (n = 6), 150 mM NaCl (n = 8); phosphate buffered 150 mM NaCl (PBS, n = 19); PBS containing 1 mg mL^{-1} GOx (n = 4); or PBS containing 5 mg mL^{-1} GOx (n = 8).

of P(AA)% achieved for KCl solutions (0.05 ± 0.01%, $n = 12$), a value indistinguishable from the no-added-electrolyte condition (0.07 ± 0.02%, $n = 28$, $p > 0.5$). It is interesting to note that the precision of these apparent permeability measurements was sufficient to reveal this subtle trend in the influence of alkali-metal cation hydrodynamic radius on P(AA)%.

Because of the logistic advantages of faster electropolymerization times, biosensors made using 1 mg mL^{-1} GOx in 150 mM KCl containing 300 mM oPD were characterized (see Table 3). Whereas enzyme loading was similar to the other designs in Table 3, the K_M value was unexpectedly high, which led to a poor LRS value. Thus, although the P(AA)% for this biosensor configuration was as good as the no-added-electrolyte condition, the S_G% value was poor because of lower substrate sensitivity. It appears, therefore, that 1 mg mL^{-1} GOx in 300 mM oPD dissolved in distilled water provided the best overall combination of good glucose sensitivity and interference (AA) rejection, yielding a biosensor with a S_G% value of 2 ± 1%.

Finally, Figure 3 (bottom) shows the trend in P(AA)% for PoPD-modified electrodes electrosynthesized in 300 mM oPD solutions for the key conditions explored in this study. The lowest (best) value was observed for the non-biosensing device prepared from the monomer dissolved in distilled water only. Addition of 1 mg mL^{-1} GOx (~5 μM) to the oPD solution doubled the subsequently determined mean P(AA)% value, whereas addition of 150 mM NaCl produced a slightly smaller detrimental effect than the much lower concentration of the macromolecular polyanion. Further addition of 40 mM phosphate ions to the 150 mM NaCl (i.e., PBS) increased P(AA)% only slightly and insignificantly. The inclusion of 1 and 5 mg mL^{-1} GOx in the PBS-based oPD solution progressively increased P(AA)%. However, the total increase in AA permeability was only a factor of four across all these conditions, so that the pure PoPD polymer ($Pt_C/PoPD_{water}$) blocked 1 mM AA by 99.94 ± 0.02% while the worst blocking was displayed by $Pt_C/PoPD_{PBS}$-GOx_5 (99.76 ± 0.04%).

The value of P(AA)% for the optimized biosensor, Pt_C/PoPD-GOx_1(H_2O), was as low as 0.11 ± 0.02% ($n = 6$; see Table 3 and Figure 3). Many literature biosensor characterization studies do not report P(AA)%, or equivalent values. However, a recent paper did report AA currents for bare and polymer-coated electrodes incorporating a novel electrosynthesized polymeric bilayer membrane composed of overoxidized poly(pyrrole) and poly(2-naphthol) films [43]. These data allow an approximate P(AA)% equivalent to be calculated: 0.13 ± 0.02%, which is not superior to the single-pot fabrication described here for Pt_C/PoPD-GOx_1(H_2O), and highlight the outstanding permselective properties of PoPD electrosynthesized under the present optimized conditions.

4. CONCLUSIONS

The analyses and results presented here demonstrate that precise measurement of PEC permeability characteristics can reveal subtle variations in the behavior of the polymer-enzyme composite layer which have important implications for biosensor design. Although variations on a common theme, all the PoPD-based biosensor designs in the present study are novel in their detail. The optimum biosensor for glucose was achieved by co-immobilizing 1 mg mL^{-1} GOx in 300 mM oPD dissolved in distilled water, a condition not reported for biosensor fabrication to date. This design showed a 3-fold superior substrate selectivity with respect to AA compared with the standard electropolymerization medium which has heretofore included an added background electrolyte, usually buffered close to neutrality. The influence of minor deviations from neutrality in these non-buffered monomer (weak base electrolyte) solutions, as well as ion-size factors, have been discussed previously for enzyme-free PoPD [42].

The improvements reported here are useful, but not critical, for glucose monitoring because of its high concentration in many body fluids. However, the approach described will help develop useful strategies in the design of biosensors for biological substrates which exist at much lower levels, such as when monitoring the key neurotransmitters, glutamate [76,77] and acetylcholine [43,78,79]. For example, the presence of enzyme protein in the polymer layer had opposite effects on permselectivity for low and high concentrations of AA, emphasizing the value of studying the concentration dependence of interference effects which is rarely reported in the literature.

Further strategies are available, such as the platinization of the smooth wire surface before PEC modification, which can increase the LRS by 60-fold and enhance BE% for a PoPD-GOx layer to ~10% [32]. A cost-benefit analysis would need to be carried out, however, before the incorporation of a further step in the fabrication process, the result of which would depend on the concentrations of the analyte and interference species in the target medium. Overall, this latest optimization of glucose biosensors based on a PoPD permselective layer, demonstrates that the one-pot fabrication of Pt_C/PoPD-GOx in the absence of added background electrolyte provides a device with AA-rejection characteristics comparable to more complex sensing layers, such as those formed from multiple electrosynthesis and over-oxidation steps [43].

ACKNOWLEDGMENTS

This work was funded in part by the Irish Research Council for Science, Engineering and Technology (IRCSET), and by Science Foundation Ireland (04/BR/C0198).

REFERENCES AND NOTES

1. Hirst, ER; Yuan, YJ; Xu, WL; Bronlund, JE. Bond-rupture immunosensors—A review. Biosens. Bioelectron **2008**, 23, 1759–1768.
2. Wanekaya, AK; Chen, W; Mulchandani, A. Recent biosensing developments in environmental security. J. Environ. Monit **2008**, 10, 703–712.
3. Sadik, OA; Aluoch, AO; Zhou, AL. Status of biomolecular recognition using electrochemical techniques. Biosens. Bioelectron **2009**, 24, 2749–2765.
4. Sozer, N; Kokini, JL. Nanotechnology and its applications in the food sector. Trends Biotechnol **2009**, 27, 82–89.
5. Saito, H; Nakazato, T; Ishii, N; Kudo, H; Otsuka, K; Endo, H; Mitsubayashi, K. An optical flow injection analysis system for measurement of glucose in tomato. Eur. Food Res. Technol **2008**, 227, 473–478.
6. Wei, D; Bailey, MJA; Andrew, P; Ryhanen, T. Electrochemical biosensors at the nanoscale. Lab Chip **2009**, 9, 2123–2131.
7. Mitsubayashi, K; Ohgoshi, T; Okamoto, T; Wakabayashi, Y; Kozuka, M; Miyajima, K; Saito, H; Kudo, H. Tonometric biosensor with a differential pressure sensor for chemo-mechanical measurement of glucose. Biosens. Bioelectron **2009**, 24, 1518–1521.
8. Lu, QZ; Lin, HL; Ge, ST; Luo, SL; Cai, QY; Grimes, CA. Wireless, remote-query, and high sensitivity escherichia coli O157:H7 biosensor based on the recognition action of Concanavalin A. Anal. Chem **2009**, 81, 5846–5850.
9. Tian, FM; Gourine, AV; Huckstepp, RTR; Dale, N. A microelectrode biosensor for real time monitoring of L-glutamate release. Anal. Chim. Acta **2009**, 645, 86–91.
10. Calia, G; Rocchitta, G; Migheli, R; Puggioni, GM; Spissu, Y; Bazzu, G; Mazzarello, V; Lowry, JP; O'Neill, RD; Desole, MS; Serra, PA. Biotelemetric monitoring of brain neurochemistry in conscious rats, using microsensors and biosensors. Sensors **2009**, 9, 2511–2523.
11. O'Neill, RD; Lowry, JP; Rocchitta, G; McMahon, CP; Serra, PA. Designing sensitive and selective polymer/enzyme composite biosensors for brain monitoring in vivo. Trends Anal. Chem **2008**, 27, 78–88.
12. Pernot, P; Mothet, JP; Schuvailo, O; Soldatkin, A; Pollegioni, L; Pilone, M; Adeline, MT; Cespuglio, R; Marinesco, S. Characterization of a yeast

D-amino acid oxidase microbiosensor for D-serine detection in the central nervous system. Anal. Chem **2008**, 80, 1589–1597.

13. Wang, J. Electrochemical glucose biosensors. Chem. Rev **2008**, 108, 814–825.
14. Wilson, GS; Gifford, R. Biosensors for real-time in vivo measurements. Biosens. Bioelectron **2005**, 20, 2388–2403.
15. Burmeister, JJ; Gerhardt, GA. Ceramic-based multisite microelectrode arrays for in vivo electrochemical recordings of glutamate and other neurochemicals. Trends Anal. Chem **2003**, 22, 498–502.
16. Pantano, P; Kuhr, WG. Enzyme-modified microelectrodes for in vivo neurochemical measurements. Electroanalysis **1995**, 7, 405–416.
17. Lowry, JP; Ryan, MR; O'Neill, RD. Behaviourally induced changes in extracellular levels of brain glutamate monitored at 1 s resolution with an implanted biosensor. Anal. Commun **1998**, 35, 87–89.
18. Schuvailo, OM; Soldatkin, OO; Lefebvre, A; Cespuglio, R; Soldatkin, AP. Highly selective microbiosensors for in vivo measurement of glucose, lactate and glutamate. Anal. Chim. Acta **2006**, 573, 110–116.
19. Dai, YQ; Zhou, DM; Shiu, KK. Permeability and permselectivity of polyphenylenediamine films synthesized at a palladium disk electrode. Electrochim. Acta **2006**, 52, 297–303.
20. Carelli, I; Chiarotto, I; Curulli, A; Palleschi, G. Electropolymerization of hydroxybenzene and aminobenzene isomers on platinum electrodes to assemble interference-free electrochemical biosensors. Electrochim. Acta **1996**, 41, 1793–1800.
21. Li, XG; Huang, MR; Duan, W; Yang, YL. Novel multifunctional polymers from aromatic diamines by oxidative polymerisations. Chem. Rev **2002**, 102, 2925–3030.
22. Killoran, SJ; O'Neill, RD. Characterization of permselective coatings electrosynthesized on Pt-Ir from the three phenylenediamine isomers for biosensor applications. Electrochim. Acta **2008**, 53, 7303–7312.
23. McMahon, CP; Killoran, SJ; O'Neill, RD. Design variations of a polymer-enzyme composite biosensor for glucose: Enhanced analyte sensitivity without increased oxygen dependence. J. Electroanal. Chem **2005**, 580, 193–202.
24. Ryan, MR; Lowry, JP; O'Neill, RD. Biosensor for neurotransmitter L-glutamic acid designed for efficient use of L-glutamate oxidase and effective rejection of interference. Analyst **1997**, 122, 1419–1424.

25. Cooper, JM; Foreman, PL; Glidle, A; Ling, TW; Pritchard, DJ. Glutamate oxidase enzyme electrodes: microsensors for neurotransmitter determination using electrochemically polymerized permselective films. J. Electroanal. Chem **1995**, 388, 143–149.

26. Sasso, SV; Pierce, RJ; Walla, R; Yacynych, AM. Electropolymerized 1, 2-diaminobenzene as a means to prevent interferences and fouling and to stabilize immobilized enzyme in electrochemical biosensors. Anal. Chem **1990**, 62, 1111–1117.

27. McMahon, CP; Rocchitta, G; Serra, PA; Kirwan, SM; Lowry, JP; O'Neill, RD. Control of the oxygen dependence of an implantable polymer/enzyme composite biosensor for glutamate. Anal. Chem **2006**, 78, 2352–2359.

28. McMahon, CP; Rocchitta, G; Kirwan, SM; Killoran, SJ; Serra, PA; Lowry, JP; O'Neill, RD. Oxygen tolerance of an implantable polymer/enzyme composite glutamate biosensor displaying polycation-enhanced substrate sensitivity. Biosens. Bioelectron **2007**, 22, 1466–1473.

29. O'Brien, KB; Killoran, SJ; O'Neill, RD; Lowry, JP. Development and characterization in vitro of a catalase-based biosensor for hydrogen peroxide monitoring. Biosens. Bioelectron **2007**, 22, 2994–3000. [Google Scholar]

30. Lowry, JP; O'Neill, RD. Partial characterization in vitro of glucose oxidase-modified poly(phenylenediamine)-coated electrodes for neurochemical analysis in vivo. Electroanalysis **1994**, 6, 369–379.

31. Malitesta, C; Palmisano, F; Torsi, L; Zambonin, PG. Glucose fast-response amperometric sensor based on glucose oxidase immobilized in an electropolymerized poly(o-phenylenediamine) film. Anal. Chem **1990**, 62, 2735–2740.

32. Reyes De Corcuera, JI; Cavalieri, RP; Powers, JR. Improved platinization conditions produce a 60-fold increase in sensitivity of amperometric biosensors using glucose oxidase immobilized in poly-o-phenylenediamine. J. Electroanal. Chem **2005**, 575, 229–241.

33. Guilbault, GG. Analytical Uses of Immobilised Enzymes; Marcel Dekker: New York, NY, USA, 1984.

34. Lee, CH; Wang, SC; Yuan, CJ; Wen, MF; Chang, KS. Comparison of amperometric biosensors fabricated by palladium sputtering, palladium electrodeposition and Nafion/carbon nanotube casting on screen-printed carbon electrodes. Biosens. Bioelectron **2007**, 22, 877–884.

35. Kulys, J; Drungiliene, A. Electrocatalytic oxidation of ascorbic acid at chemically modified electrodes. Electroanalysis **1991**, 3, 209–214.

36. El Atrash, SS; O'Neill, RD. Characterisation in vitro of a naphthoquinone-mediated glucose oxidase-modified carbon paste electrode designed for neurochemical analysis in vivo. Electrochim. Acta **1995**, 40, 2791–2797.

37. Lin, YQ; Liu, K; Yu, P; Xiang, L; Li, XC; Mao, LQ. A facille electrochemical method for simultaneous and on-line measurements of glucose and lactate in brain microdialysate with prussian blue as the electrocatalyst for reduction of hydrogen peroxide. Anal. Chem **2007**, 79, 9577–9583.

38. Fu, YC; Chen, C; Xie, QJ; Xu, XH; Zou, C; Zhou, QM; Tan, L; Tang, H; Zhang, YY; Yao, SZ. Immobilization of enzymes through one-pot chemical preoxidation and electropolymerization of dithiols in enzyme-containing aqueous suspensions to develop biosensors with improved performance. Anal. Chem **2008**, 80, 5829–5838.

39. Lowry, JP; McAteer, K; El Atrash, SS; Duff, A; O'Neill, RD. Characterization of glucose oxidase-modified poly(phenylenediamine)-coated electrodes in vitro and in vivo: Homogeneous interference by ascorbic acid in hydrogen peroxide detection. Anal. Chem **1994**, 66, 1754–1761.

40. Wang, J; Chen, L; Liu, J; Lu, F. Enhanced selectivity and sensitivity of first-generation enzyme electrodes based on the coupling of rhodinized carbon paste transducers and permselective poly(o-phenylenediamine) coatings. Electroanalysis **1996**, 8, 1127–1130.

41. Rothwell, SA; Kinsella, ME; Zain, ZM; Serra, PA; Rocchitta, G; Lowry, JP; O'Neill, RD. Contributions by a novel edge effect to the permselectivity of an electrosynthesized polymer for microbiosensor applications. Anal. Chem **2009**, 81, 3911–3918.

42. Rothwell, SA; Killoran, SJ; Neville, EM; Crotty, AM; O'Neill, RD. Poly(o-phenylenediamine) electrosynthesized in the absence of added background electrolyte provides a new permselectivity benchmark for biosensor applications. Electrochem. Commun **2008**, 10, 1078–1081. [Google Scholar]

43. Guerrieri, A; Lattanzio, V; Palmisano, F; Zambonin, PG. Electrosynthesized poly(pyrrole)/poly(2-naphthol) bilayer membrane as an effective anti-interference layer for simultaneous determination of acethylcholine and choline by a dual electrode amperometric biosensor. Biosens. Bioelectron **2006**, 21, 1710–1718.

44. Kirwan, SM; Rocchitta, G; McMahon, CP; Craig, JD; Killoran, SJ; O'Brien, KB; Serra, PA; Lowry, JP; O'Neill, RD. Modifications of poly(o-phenylenediamine) permselective layer on Pt-Ir for biosensor application in neurochemical monitoring. Sensors **2007**, 7, 420–437.

45. Craig, JD; O'Neill, RD. Comparison of simple aromatic amines for electrosynthesis of permselective polymers in biosensor fabrication. Analyst **2003**, 128, 905–911.

46. Myler, S; Eaton, S; Higson, SPJ. Poly(o-phenylenediamine) ultra-thin polymer-film composite membranes for enzyme electrodes. Anal. Chim. Acta **1997**, 357, 55–61.

47. Gooding, JJ; Hall, EAH. Parameters in the design of oxygen detecting oxidase enzyme electrodes. Electroanalysis **1996**, 8, 407–413.

48. Dixon, BM; Lowry, JP; O'Neill, RD. Characterization in vitro and in vivo of the oxygen dependence of an enzyme/polymer biosensor for monitoring brain glucose. J. Neurosci. Meth **2002**, 119, 135–142.

49. Compagnone, D; Federici, G; Bannister, JV. A new conducting polymer glucose sensor based on polythianaphthene. Electroanalysis **1996**, 7, 1151–1155.

50. Centonze, D; Guerrieri, A; Malitesta, C; Palmisano, F; Zambonin, PG. An in-situ electrosynthesized poly-ortho-phenylenediamine/glucose oxidase amperometric biosensor for flow-injection determination of glucose in serum. Ann. Chim. (Rome) **1992**, 82, 219–234.

51. Miele, M; Fillenz, M. In vivo determination of extracellular brain ascorbate. J. Neurosci. Meth **1996**, 70, 15–19.

52. Boutelle, MG; Svensson, L; Fillenz, M. Rapid changes in striatal ascorbate in response to tail-pinch monitored by constant potential voltammetry. Neuroscience **1989**, 30, 11–17.

53. O'Neill, RD; Fillenz, M; Albery, WJ. The development of linear sweep voltammetry with carbon paste electrodes in vivo. J. Neurosci. Meth **1983**, 8, 263–273.

54. Fillenz, M; O'Neill, RD. Effects of light reversal on the circadian pattern of motor activity and voltammetric signals recorded in rat forebrain. J. Physiol. (London) **1986**, 374, 91–101.

55. Brose, N; O'Neill, RD; Boutelle, MG; Anderson, SMP; Fillenz, M. Effects of an anxiogenic benzodiazepine receptor ligand on rat motor activity and dopamine release in nucleus accumbens and striatum. J. Neurosci **1987**, 7, 2917–2926.

56. Soldatkin, OO; Schuvailo, OM; Marinesco, S; Cespuglio, R; Soldatkin, AR. Microbiosensor based on glucose oxidase and hexokinase co-immobilised on platinum microelectrode for selective ATP detection. Talanta **2009**, 78, 1023–1028.

57. Santos, RM; Lourenco, CF; Piedade, AP; Andrews, R; Pomerleau, F; Huettl, P; Gerhardt, GA; Laranjinha, J; Barbosa, RM. A comparative study of carbon fiber-based microelectrodes for the measurement of nitric oxide in brain tissue. Biosens. Bioelectron **2008**, 24, 704–709.

58. Hamdi, N; Wang, JJ; Monbouquette, HG. Polymer films as permselective coatings for H_2O_2-sensing electrodes. J. Electroanal. Chem **2005**, 581, 258–264.

59. McAteer, K; O'Neill, RD. Strategies for decreasing ascorbate interference at glucose oxidase-modified poly(o-phenylenediamine)-coated electrodes. Analyst **1996**, 121, 773–777.

60. Lowry, JP; O'Neill, RD; Boutelle, MG; Fillenz, M. Continuous monitoring of extracellular glucose concentrations in the striatum of freely moving rats with an implanted glucose biosensor. J. Neurochem **1998**, 70, 391–396.

61. Dai, YQ; Shiu, KK. Highly sensitive amperometric glucose biosensor based on glassy carbon electrode with copper/palladium coating. Electroanalysis **2004**, 16, 1806–1813.

62. Bartlett, PN; Wang, JH; James, W. Measurement of low glucose concentrations using a microelectrochemical enzyme transistor. Analyst **1998**, 123, 387–392.

63. Wang, J; Lu, F. Oxygen-rich oxidase enzyme electrodes for operation in oxygen-free solutions. J. Am. Chem. Soc **1998**, 120, 1048–1050.

64. Wang, J. Selectivity coefficients for amperometric sensors. Talanta **1994**, 41, 857–863.

65. Palmisano, F; Rizzi, R; Centonze, D; Zambonin, PG. Simultaneous monitoring of glucose and lactate by an interference and cross-talk free dual electrode amperometric biosensor based on electropolymerized thin films. Biosens. Bioelectron **2000**, 15, 531–539.

66. Ahmad, F; Christenson, A; Bainbridge, M; Yusof, APM; Ab Ghani, S. Minimizing tissue-material interaction in microsensor for subcutaneous glucose monitoring. Biosens. Bioelectron **2007**, 22, 1625–1632.

67. Maalouf, R; Chebib, H; Saikali, Y; Vittori, O; Sigaud, M; Garrelie, F; Donnet, C; Jaffirezic-Renault, N. Characterization of different diamond-

like carbon electrodes for biosensor design. Talanta **2007**, 72, 310–314. [Google Scholar]

68. Cooper, JM; Pritchard, DJ. Biomolecular sensors for neurotransmitter determination electrochemical immobilization of glutamate oxidase at microelectrodes in a poly (o-phenylenediamine) film. J. Mater. Sci.: Mater. Electronl **1994**, 5, 111–116.

69. Bao, L; Avshalumov, MV; Patel, JC; Lee, CR; Miller, EW; Chang, CJ; Rice, ME. Mitochondria are the source of hydrogen peroxide for dynamic brain-cell signaling. J. Neurosci **2009**, 29, 9002–9010.

70. McMahon, CP; Rocchitta, G; Serra, PA; Kirwan, SM; Lowry, JP; O'Neill, RD. The efficiency of immobilised glutamate oxidase decreases with surface enzyme loading: an electrostatic effect, and reversal by a polycation significantly enhances biosensor sensitivity. Analyst **2006**, 131, 68–72.

71. Burmeister, JJ; Palmer, M; Gerhardt, GA. Ceramic-based multisite microelectrode array for rapid choline measures in brain tissue. Anal. Chim. Acta **2003**, 481, 65–74.

72. Berners, MOM; Boutelle, MG; Fillenz, M. On-line measurement of brain glutamate with an enzyme/polymer-coated tubular electrode. Anal. Chem **1994**, 66, 2017–2021.

73. Kulagina, NV; Shankar, L; Michael, AC. Monitoring glutamate and ascorbate in the extracellular space of brain tissue with electrochemical microsensors. Anal. Chem **1999**, 71, 5093–5100.

74. Harrar, JE. Controlled-potential coulometric determination of hydrogen peroxide. Anal. Chem **1963**, 35, 893–896.

75. Rothwell, SA; McMahon, CP; O'Neill, RD. Effects of polymerization potential on the permselectivity of poly(o-phenylenediamine) coatings deposited on Pt-Ir electrodes for biosensor applications. Electrochim. Acta **2010**, 55, 1051–1060.

76. Hascup, KN; Hascup, ER; Pomerleau, F; Huettl, P; Gerhardt, GA. Second-by-second measures of L-glutamate in the prefrontal cortex and striatum of freely moving mice. J. Pharmacol. Exp. Ther **2008**, 324, 725–731.

77. Morales-Villagran, A; Medina-Ceja, L; Lopez-Perez, SJ. Simultaneous glutamate and EEG activity measurements during seizures in rat hippocampal region with the use of an electrochemical biosensor. J. Neurosci. Meth **2008**, 168, 48–53.

78. Du, D; Ding, JW; Cai, J; Zhang, AD. One-step electrochemically deposited interface of chitosan-gold nanoparticles for

acetylcholinesterase biosensor design. J. Electroanal. Chem **2007**, 605, 53–60.

79. Dale, N; Hatz, S; Tian, FM; Llaudet, E. Listening to the brain: microelectrode biosensors for neurochemicals. Trends Biotechnol **2005**, 23, 420–428.

80. Patel, BA; Arundell, M; Parker, KH; Yeoman, MS; O'Hare, D. Detection of nitric oxide release from single neurons in the pond snail, Lymnaea stagnalis. Anal. Chem **2006**, 78, 7643–7648.

81. Losito, I; Palmisano, F; Zambonin, PG. o-Phenylenediamine electropolymerization by cyclic voltammetry combined with electrospray ionization-ion trap mass spectrometry. Anal. Chem **2003**, 75, 4988–4995.

82. Palmisano, F; Zambonin, PG; Centonze, D. Amperometric biosensors based on electrosynthesised polymeric films. Fresenius J. Anal. Chem **2000**, 366, 586–601.

83. Centonze, D; Losito, I; Malitesta, C; Palmisano, F; Zambonin, PG. Electrochemical immobilisation of enzymes on conducting organic salt electrodes: characterisation of an oxygen independent and interference-free glucose biosensor. J. Electroanal. Chem **1997**, 435, 103–111.

84. Wang, Q; Tang, H; Xie, QJ; Jia, XE; Zhang, YY; Tan, L; Yao, SZ. The preparation and characterization of poly(o-phenylenediamine)/gold nanoparticles interface for immunoassay by surface plasmon resonance and electrochemistry. Colloid. Surface. B **2008**, 63, 254–261.

85. Camurri, G; Ferrarini, P; Giovanardi, R; Benassi, R; Fontanesi, C. Modelling of the initial stages of the electropolymerization mechanism of o-phenylenediamine. J. Electroanal. Chem **2005**, 585, 181–190. [Google Scholar]

86. Losito, I; De Giglio, E; Cioffi, N; Malitesta, C. Spectroscopic investigation on polymer films obtained by oxidation of o-phenylenediamine on platinum electrodes at different pHs. J. Mater. Chem **2001**, 11, 1812–1817.

87. Ekinci, E; Erdogdu, G; Karagozler, AE. Preparation, optimization, and voltammetric characteristics of poly(o-phenylenediamine) film as a dopamine-selective polymeric membrane. J. Appl. Polym. Sci **2001**, 79, 327–332.

88. Centonze, D; Malitesta, C; Palmisano, F; Zambonin, PG. Permeation of solutes through an electropolymerized ultrathin poly-o-

phenylenediamine film used as an enzyme-entrapping membrane. Electroanalysis **1994**, 6, 423–429.

Chapter 8

Vertically Aligned Zno Nanorod Core-Polypyrrole Conducting Polymer Sheath And Nanotube Arrays For Electrochemical Supercapacitor Energy Storage

Navjot Kaur Sidhu and Alok C Rastogi

Electrical and Computer Engineering Department and Center for Autonomous Solar Power (CASP), Binghamton University, State University of New York, New York 13902, USA

ABSTRACT

Nanocomposite electrodes having three-dimensional (3-D) nanoscale architecture comprising of vertically aligned ZnO nanorod array core-polypyrrole (PPy) conducting polymer sheath and the vertical PPy nanotube arrays have been investigated for supercapacitor energy storage. The electrodes in the ZnO nanorod core-PPy sheath structure are formed by preferential nucleation and deposition of PPy layer over hydrothermally synthesized vertical ZnO nanorod array by controlled pulsed current electropolymerization of pyrrole monomer under surfactant action. The vertical PPy nanotube arrays of different tube diameter are created by selective etching of the ZnO nanorod core in ammonia solution for different periods. Cyclic voltammetry studies show high areal-specific capacitance approximately 240 $mF.cm^{-2}$ for open pore and approximately 180 $mF.cm^{-2}$ for narrow 30-to-36-nm diameter PPy nanotube arrays attributed to intensive faradic processes arising from enhanced access of electrolyte ions through nanotube interior and exterior. Impedance spectroscopy studies show that capacitive response extends over larger frequency domain in electrodes with PPy nanotube structure. Simulation of Nyquist plots by electrical equivalent circuit modeling establishes that 3-D nanostructure is better represented by constant phase element which accounts for the inhomogeneous electrochemical redox processes. Charge-discharge studies at different current densities establish that kinetics of the redox process in PPy nanotube electrode is due to the limitation on electron transport rather than the diffusive process of electrolyte ions. The PPy nanotube electrodes show deep discharge capability with high coulomb efficiency and long-term charge-discharge cyclic studies show nondegrading performance of the specific areal capacitance tested for 5,000 cycles.

Keywords Zinc oxide nanorods Polypyrrole nanotubes 3-D nanostructures Supercapacitor Pulsed electrochemical polymerization Electrochemical energy storage Redox capacitance

BACKGROUND

Electrochemical energy storage in the ultracapacitor devices is emerging as a frontline technology for high-power applications ranging from modern portable electronics to electric automotive. A battery-supercapacitor hybrid energy system is a power source that can meet the peak power demands in camera flashes, pulsed lasers, and computer systems back-up as well as electric propulsion in diverse industrial and vehicular transport applications. Among the materials systems, structured carbons which store charges as an electric double layer (EDL) in liquid electrolyte medium are widely studied with a focus on overcoming the energy-density limitation [1]. The materials systems which show capacitive function based on redox reactions are the insertion-type metal oxides and doped-conducting polymers capable of high energy-density storage [2, 3]. The conducting polymers, such as polypyrrole (PPy), poly(3,4 ethylenedioxythiophene) (PEDOT), and polyaniline (PANI) which undergo redox processes equivalent of doping and dedoping of electrolyte ions as means of energy storage are being aggressively studied. These polymers exhibit pseudocapacitance properties due to presence of charge transfer reactions. The other most widely studied materials are the metal oxides RuO_2, MnO_2, V_2O_5, NiO, and Co_3O_4 which show highly capacitive behavior due to reversible and fast surface redox reactions with electrolyte ions [2, 4].

In the recent years, conducting polymers with a nanoporous morphology and as nanocomposites with metal-oxides have emerged as the materials system of great potential for high energy-density storage. Electrodes based on these materials structured at the nanoscale enable many-fold enhancements of the electroactive surface and interface with electrolyte facilitating absorption, ingress, and diffusion of electrolyte ions which being the main energy storage units could lead to increased energy and power density of supercapacitor devices. The high surface area morphology in conducting polymers is attained by creating variations in its nanostructure like nanoporous [5], nanofibers [6, 7], nanowires [8], nanobelts [9], and by size-selective nanopores in the context of carbons [10]. Most metal oxides are electrically resistive in character and the redox reactions here are limited to the surface regions. In order to accrue full advantage of their pseudocapacitive function, electrically conducting polypyrrole is encrusted by metal oxides [11, 12], PEDOT [13], CNT [14], and graphene [15, 16] in various nanostructured forms like nanoparticle [17, 18], nanotube [19, 20], and nanowire [21, 22] as supercapacitor electrode. In these materials systems, the nanostructure features are randomly distributed in the two-dimensional (2-D) film form mainly due to the preparatory methods.

Most recent research thrust in the conducting polymers and their nanocomposite with metal oxides is directed towards the electrodes with three-dimensional (3-D) nanoarchitecture such as vertically aligned nanotubes [23] and nanorods [24]. These nanostructures have potential for the limiting electrolyte-ion diffusion problem by decreasing the ion diffusion paths and at the same time increasing the surface area for enhanced electrode-electrolyte interaction. In the past, randomly oriented conducting polymer nanotubes structures have been synthesized [16, 25, 26] for supercapacitor applications. However, the vertically oriented nanostructures, nanorods, and nanotubes have been mostly configured using the metal oxide templates [27]. Such nanostructures have been created by more innovating nanoscale engineering methods like oxidative polymerization [28], electrochemical anodic oxidation [29], electrodeposition [30], and hydrothermal synthesis [31, 32]. Furthermore, by combining the redox conducting polymers with the well-known pseudocapacitive oxide like MnO_2, forming the nanocomposites in the 3-D nanoarchitecture presents multiple advantages with enormous potential to outperform their 2-D counterparts. The composite 3-D nanostructure can be created by conformal deposition of redox-active conducting polymer, pseudocapacitive oxide layer, or their multilayer stacks over vertical nanostructures of TiO_2, ZnO, or NiO serving as templates. The composite 3-D nanostructured electrodes have synergic contribution to specific capacitance based on their electroactive functions which boost energy density, and their nanoarchitecture have the ability to mitigate the ion diffusion limitation thereby enhancing the power density. In the past, 3-D nanotube polymers, PPy-PANI [33] polymer-metal oxides, TiO_2-PPy [34, 35], ZnO-PPy [36], TiO_2-NiO [23], and TiO_2-V_2O_5[37] have been reported.

In this work, we investigate the characteristics of nanocomposite electrodes for supercapacitors having 3-D nanoscale architecture, the one comprising of vertically aligned zinc oxide nanorod arrays at the core with doped-polypyrrole conducting polymer sheath and the other vertical polypyrrole nanotubes arrays. Although polypyrrole in the doped state shows high electrical conductivity, the conversion between redox states is very slow due to the slow transportation of counter ions to balance the charge in the polymer structure

[38]. The vertical polypyrrole nanotube and sheath structure are likely to decrease the charge transfer reaction time and thus enhance the charge storage capabilities [38]. Zinc oxide used prominently in various devices such as biosensors [39], light emitting diodes [40], organic solar cells [41], and spintronics [42, 43] is a biocompatible, highly stable, and less expensive material as compared to ruthenium oxide and therefore has a good potential for the electrochemical energy storage devices. Zinc oxide characterized as a wide band gap semiconductor with excellent chemical and physical properties can be easily transformed in various nanostructure forms like nanowire, nanoplatelets, and nanoneedles mostly as flat two-dimensional structures [44]. In the context of using a ZnO template for a supercapacitor electrode in the 3-D architecture, we have fabricated vertically aligned ZnO nanorods by hydrothermal synthesis which exhibit specific electrical and optical properties [32]. The nanocomposite electrode is formed by deposition of doped polypyrrole layer over ZnO nanorod at the core by the commercially viable, low cost solution-based pulsed current electropolymerization process [45]. Pulsed current process allows depositing polypyrrole layer selectively with highly controlled thickness through the application of number of pulses. Only a few studies have been reported on electrodes with zinc oxide and polypyrrole composite films for supercapacitor energy storage devices perhaps since zinc oxide nanostructures may be resistive compared to conducting polymers [46]. ZnO nanorods as template to create PPy nanotube structures with inlay of MnO_2 and their energy storage behavior have been reported [36]. In this work, the conducting doped polypyrrole supercapacitor electrodes in two different 3-D architectural forms, one having ZnO nanorod core-PPy sheath and the other vertical PPy nanotube array have been investigated. The electrochemical properties of these electrodes were studied by impedance spectroscopy, cyclic voltammetry, and charge-discharge measurements. Randles circuit model with additional capacitance and resistance elements is developed to explain the characteristics of electrode at various frequency ranges. Long-term charge-discharge tests are carried out to evaluate the cycle life of such electrodes in supercapacitor energy storage device. This paper reports the results of these studies.

METHODS

Synthesis of ZnO nanorod array template

Polypyrrole conducting polymer electrodes in the two ZnO core-PPy sheath and PPy nanotube structural forms were fabricated over a ZnO nanorod template. The template, a vertically oriented two-dimensional array of ZnO nanorods, was formed over surface-activated 500-µm-thin graphite substrates by thermo-decomposition of zinc nitrate aqueous solution in the presence of hexamethylenetetramine in a wet chemical process [32]. The surface of the graphite substrate is activated by depositing a 20-nm-thick ZnO seed layer that acts as nucleation centers for the growth of ZnO nanorods. This layer is formed by radio-frequency (RF) plasma sputtering from a stoichiometric ZnO target in the argon ambient at 50 mTorr pressure and 100-W RF power. The growth of ZnO nanorods is done in a solution of 0.03 M zinc nitrate hexahydrate (Sigma-Aldrich) and equimolar hexamethylenetetramine (HMT) (Sigma-Aldrich, St. Louis, MO, USA) in 18 MΩ deionized water. The surface-activated graphite substrate is vertically kept in this solution in the autoclavable glass bottle and held steady at 95°C for 8 hours. This hydrothermal procedure results in the dissociation of Zn(II)-amino complex resulting in ZnO which grows as ZnO nanorods [47]. Post deposition, the graphite substrates are rinsed in deionized water to remove any residual precursor chemicals and dried in air.

Pulsed current electropolymerization of polypyrrole sheath and nanotube formation

A uniform and conformal deposition of polypyrrole of a controlled thickness over ZnO nanorods is essential for the creation of PPy sheath and nanotube nanostructures. Polypyrrole has been deposited the past by various chemical [48] and potentiostatic electropolymerization [49] methods. In this work, PPy deposition is done by electropolymerization of pyrrole monomer in situ over ZnO nanorods using the ultrashort multiple unipolar pulsed-current method reported earlier [45]. Electropolymerization is accomplished in an electrolyte solution containing 0.01 M pyrrole (Py) monomer (Sigma-Aldrich) and 0.1 M lithium perchlorate dopant ions in the presence of 0.06 M sodium dodecyl sulfate (SDS) surfactant. The preparatory steps involve first the dissolution of pyrrole monomer in the presence of SDS under continuous stirring in deionized water warmed to 40°C. Later, the ZnO-seed-layer-coated graphite substrate is dipped vertically for approximately 20 min in this solution which helps in the wettability of the ZnO nanorods with pyrrole monomer. Before initiating the electropolymerization process, lithium perchlorate is added to form 0.1 M solution. The formation of the polypyrrole sheath layer over ZnO nanorods is carried out by pulsed current electropolymerization method in a two-electrode cell using a platinum (Pt) sheet as a counter and a reference electrode. In this method, multiple unipolar anodic ultrashort 10-ms-duration constant-current pulses of amplitude 4 $mA.cm^{-2}$ are applied. Each of the pulses is interspaced by a current 'off' period of 100-ms duration. The electropolymerization is initiated during the pulse 'on' period as the corresponding anodic potential exceeds the oxidation potential of the pyrrole monomer. The current off period essentially helps create the equilibrium conditions in the vicinity of ZnO nanorods for deposition under homogenous polymerization conditions. The number of current pulses effectively controls the polypyrrole-layer thickness and usually approximately 5 to 10 k pulses were used to form fully covered PPy sheath over

ZnO nanorods. In some cases 20 k pulses were also applied to form a thicker PPy sheath.

The PPy nanotube structure is obtained by etching away the vertically aligned ZnO nanorod core in a 20% ammonia solution [36]. The ZnO etching process is typically initiated over ZnO nanorod tips due to relatively thin cladding of the electropolymerized polypyrrole. As a result, the PPy nanotube structure shows dependence on the etching time. In this work, etching times of 2 and 4 h are used for the formation of PPy nanotube arrays.

Electrochemical characterization of supercapacitor electrodes

Efficacy of the ZnO nanorod core-PPy sheath and PPy nanotube electrodes for the supercapacitor energy storage device application was analyzed by various electrochemical characterizations. These electrodes were characterized by cyclic voltammetry (CV), alternating current (ac) impedance spectroscopy, galvanic charge-discharge, and long-term cyclic tests in a three-electrode cell with Pt sheet as counter electrode and the potential referenced to a saturated Ag/AgCl electrode in an electrolyte comprising of an aqueous solution of 1 M lithium perchlorate. Cyclic voltammetry and galvanic charge-discharge measurements were carried out using Solartron electrochemical interface (Model 1287 from Solartron Analytical, Oak Ridge, TN, USA). In cyclic voltammetry, the flow of electric current between the working electrode and Pt counter electrode was recorded in the potential range -0.5 to +0.5 V scanned at different rates between 5 to100 $mV.s^{-1}$. The areal-specific capacitance, C_{sv} ($F.cm^{-2}$), of the electrodes was calculated using the relation,

$$C_{sv} = \frac{i_a + |i_c|}{2s} \tag{1}$$

where i_a and i_c are the absolute values of the anodic and cathodic current ($mA.cm^{-2}$) of the electrode area and s is the scan rate ($mV.s^{-1}$). The galvanic charge-discharge characteristics were measured at various current densities, i_d, varying between 1, 2, and 3 $mA.cm^{-2}$ in the potential range of 0.05 to 0.5 V. In the discharge cycle, using the discharge time, Δt, and a corresponding change in voltage, ΔV, excluding the IR voltage drop, the areal-specific capacitance C_{sd} ($F.cm^{-2}$) is calculated by the relation,

$$C_{sd} = \frac{i_d \cdot \Delta t}{\Delta V} \tag{2}$$

The ac impedance measurements were carried out in a two-electrode configuration in the frequency range 1 mHz to 100 kHz with ac signal amplitude of 10 mV using Solartron Impedance/Gain-Phase Analyzer (Model 1260).

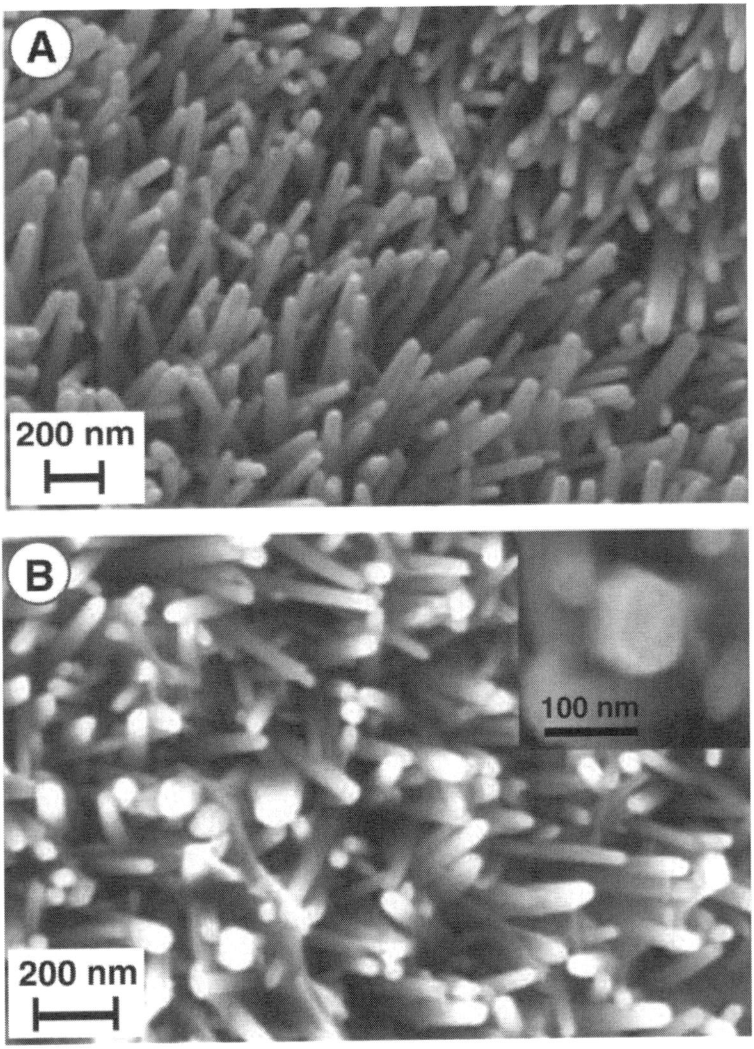

Figure 1 SEM images of ZnO nanorod arrays grown on graphite substrate. (A) Image showing the microstructure of ZnO nanorod arrays. (B) Magnified image showing the top end of the rods with hexagonal facets.

Measured low-frequency imaginary impedance Z'' provides estimate of the overall capacitance C_i using the relation $C_i = 1/|\omega Z''|$. The Nyquist plots using the impedance data were simulated using the equivalent electrical model representing the electrochemical and electrophysical attributes of the nanostructured ZnO-PPy electrode using ZPlot software (Scribner Associates, NC, USA) which provide the characteristic resistances and various contributing factors to the overall electrode capacitance.

RESULTS AND DISCUSSION

Microstructure of ZnO nanorod core-polypyrrole sheath, nanotube electrodes

The microstructure of ZnO nanorod arrays grown over graphite substrates is shown by SEM micrograph in Figure 1A. These vertically grown ZnO nanorods are homogeneously dispersed across the substrate surface and their average length dependent on the growth time is typically approximately 2.2 to 2.5 μm. These rods appear to have smooth surface texture and nearly uniform average diameter of approximately 60 nm. As shown in the magnified image in Figure 1B and its inset, the top end of these rods have a hexagonal facet signifying these rods grow along the crystalline *c*-axis.

In the formation of PPy sheath over ZnO nanorods, its thickness is controlled by the number of pulsed current cycles. Figure 2A shows the early steps of the pulsed polymerization representing the formative stages of the growth of polypyrrole layer over ZnO nanorod arrays. It shows that the polypyrrole layer consisting of small compact nodular features forms conformal to the ZnO nanorods across its entire length. The nodular surface structure of polypyrrole layer is due to congregation of pyrrole monomer resulting from the action of SDS surfactant [50]. Furthermore, there is no deposition of polypyrrole in the interrod space and the PPy sheath forms preferentially over ZnO nanorods due to pyrrole monomer incursion by the action of the SDS surfactant as discussed later [50]. The inset shows a magnified view of a ZnO nanorod at the core coated with PPy sheath having overall average diameter of approximately 110 nm. Figure 2B shows ZnO core-PPy shell structure after electropolymerization has been accomplished for the full 10 k unipolar pulsed current cycles. The average diameter of the ZnO-core-PPy shell grows to approximately 360 nm which translates to approximately 150 nm average thickness of the PPy layer as shown by the magnified view of the top of ZnO nanorods in the inset of Figure 2B. At this growth stage, the inter-ZnO nanorod space begins to fill due to the coalescence of PPy sheath formed over different ZnO nanorods in the array. For the creation of the freestanding PPy nanotube array, the ZnO nanorod in the core is etched away in 20% ammonia solution. Figure 2C shows the partial etched state of the ZnO core for 2 h which creates tubular holes of approximately 30 to 36 nm in average diameter as shown in the inset of Figure 2C. At this stage, the PPy nanotube arrays still have in their interior a finite thickness of ZnO cladding. To remove the ZnO cladding, additional etching was carried out. It was observed that after a prolonged etching for approximately 4 h, a complete removal of the ZnO cladding was realized which resulted in the formation of a network of PPy nanotubular arrays as shown in the micrograph in Figure 2D. A magnified view in the inset shows PPy

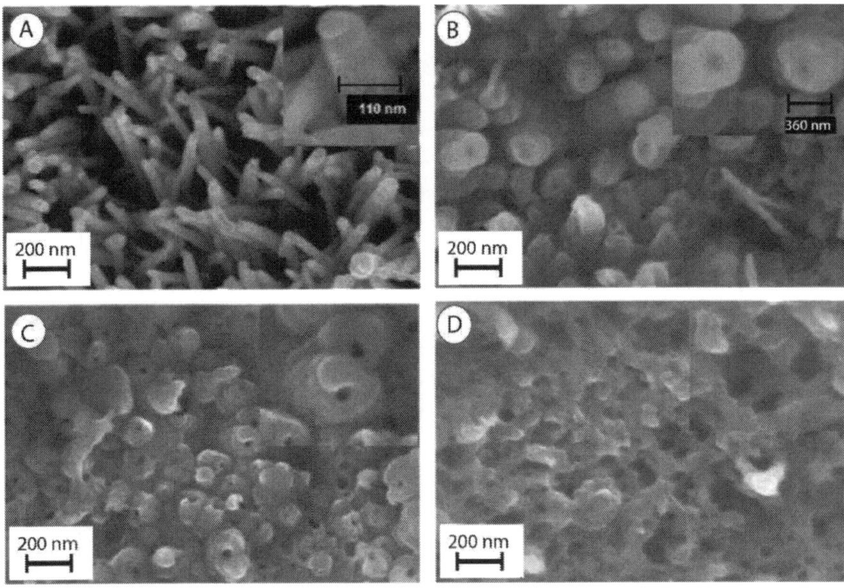

Figure 2 SEM images of ZnO nanorod arrays coated with pulsed current polymerized PPy sheath. (A) Initial stage of PPy oligomers cluster deposition, **(B)** ZnO core-PPy sheath structure after 10 k pulsed electropolymerization cycles, **(C)** PPy nanotube array after 2-h etch, and (D) open pore PPy nanotube array after 4-h etch.

nanotubes of diameter approximately 60 to 70 nm consistent with the typical diameter of the ZnO nanorod core. Figure 2D also shows that a large number of these PPy nanotubes share a common sheath wall which had initially resulted from the PPy growth in the space between neighboring ZnO nanorods. This enables formation of the interconnected open network of freestanding PPy nanotube arrays. The PPy nanotube diameter can be enhanced by forming thicker ZnO nanorod array core structure. However, this reduces the effective thickness of PPy tubular sheath and hence the effective mass of PPy which is an active component for charge storage. On the other hand, increasing thickness of PPy by electropolymerization for longer pulsed current cycles excessively covers the top of the ZnO nanorod arrays making it difficult to etch away the ZnO core which prevented realization of PPy nanotubular arrays. Figure 3 shows the ZnO core-PPy sheath structure with the thicker PPy layer deposited using 20 k unipolar pulsed current cycles. This results in formation of thick conjoined PPy sheath with thickly deposited PPy over the top of ZnO nanorods (Figure 3A). Figure 3B shows a cross-sectional view indicating the ZnO nanorods could still be coated with PPy along its length. The side panel in Figure 3C shows conjoined PPy sheath over ZnO nanorods of average diameter approximately 985 nm to 1 μm. Morphology of the thick PPy deposit is like nodules. Figure 3D shows the top view of the PPy coated ZnO nanorods tips. Figure 3E shows the same view after ammonia etching for 4 h. It is evident that such ZnO nanorod core-PPy sheath structure did not result in the PPy nanotube structure after etching. The evolution of the PPy sheath and nanotube structure is schematically shown in Figures 4A, B, C, D, E, F. The vertical ZnO nanorod array (Figure 4A) is preferentially coated with PPy by pulsed electropolymerization process through surfactant action. Progressively, on continued pulsed current polymerization cycles, the PPy sheath thickness increases (Figure 4B) with possible merging of PPy sheath walls (Figure 4C). Figures 4D, E, F show the evolution of PPy nanotubes through etching of ZnO core starting at the nanorod tips which after short term etching results in the PPy nanotubes along with the inverted conical ZnO cladding (Figure 4D). The PPy nanotube arrays without the ZnO cladding are created by complete etching of ZnO for longer periods as depicted in Figure 4E with an open pore structure as shown in the top view in Figure 4F.

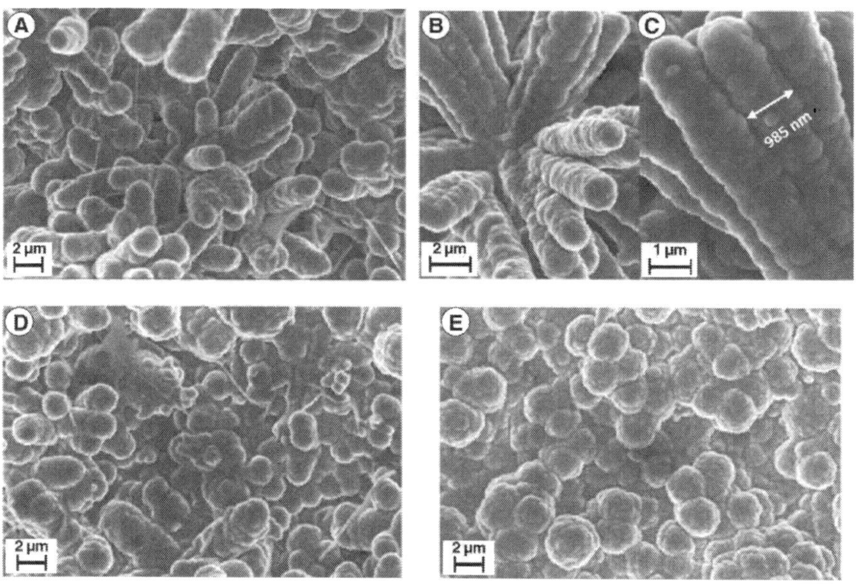

Figure 3 SEM images. (A) Thicker PPy deposited over ZnO nanorod array when electropolymerization was carried out for 20 k pulsed current cycles, (B) cross-sectional view of PPy sheath coated along the ZnO rod length, and **(C)** conjoined view of PPy sheath over ZnO nanorods with average diameter of 985 nm. Top view of ZnO nanorod tips with thick PPy sheath (D) before etch and (E) after ammonia etch.

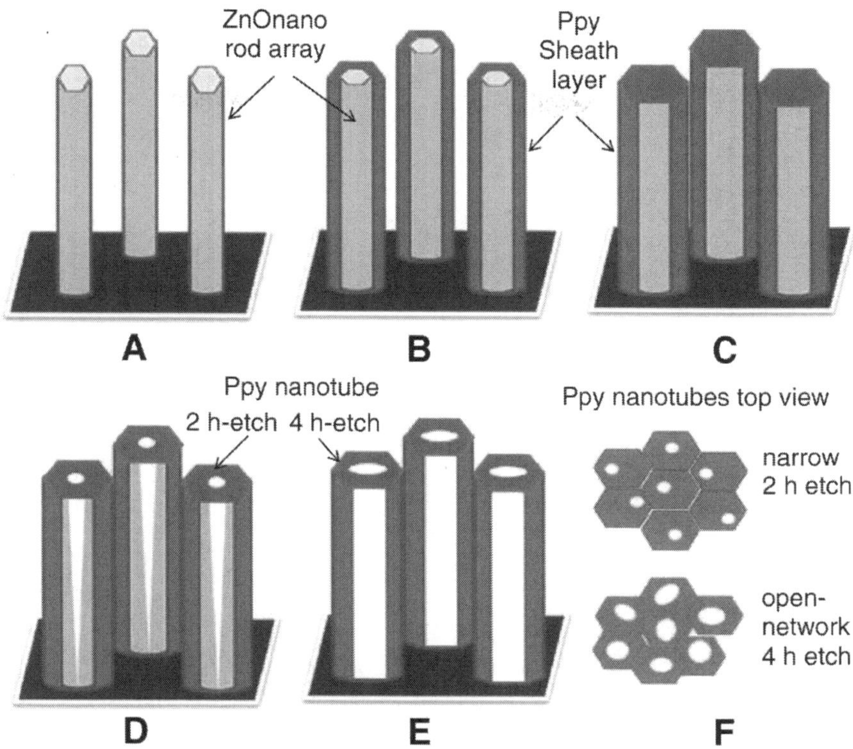

Figure 4 Schematic representation of the evolution of the PPy sheath and nanotube structure. Schematic representation of ZnO nanorod core coated by PPy sheath **(A-C)** and formation of PPy nanotube array after 2- and 4-h etch **(D-E)**. Top view **(F)**.

Growth features of ZnO nanorod-PPy sheath and PPy nanotube arrays

Unlike the two-dimensional flat conducting substrates in which case conventional direct current (dc) potentiostatic electropolymerization of pyrrole can produce uniform thick polypyrrole film, over the semiconducting ZnO nanostructures, pulsed current electropolymerization employed in this work was found essential to obtain homogeneous polypyrrole sheath. In order to create PPy 3-D tubular nanostructures for energy storage action, it is essential (i) to form the PPy sheath in the high-conductivity anion doped state and (ii) to have the PPy sheath of desired thickness coated uniformly over the entire length of the ZnO nanorod array at its core. The first criterion is largely met by anodic electropolymerization of pyrrole monomer in the aqueous medium in the presence of ClO_4^{2-} anions derived from $LiClO_4$ in the electrolyte. Various mechanisms of pyrrole electropolymerization have been proposed under

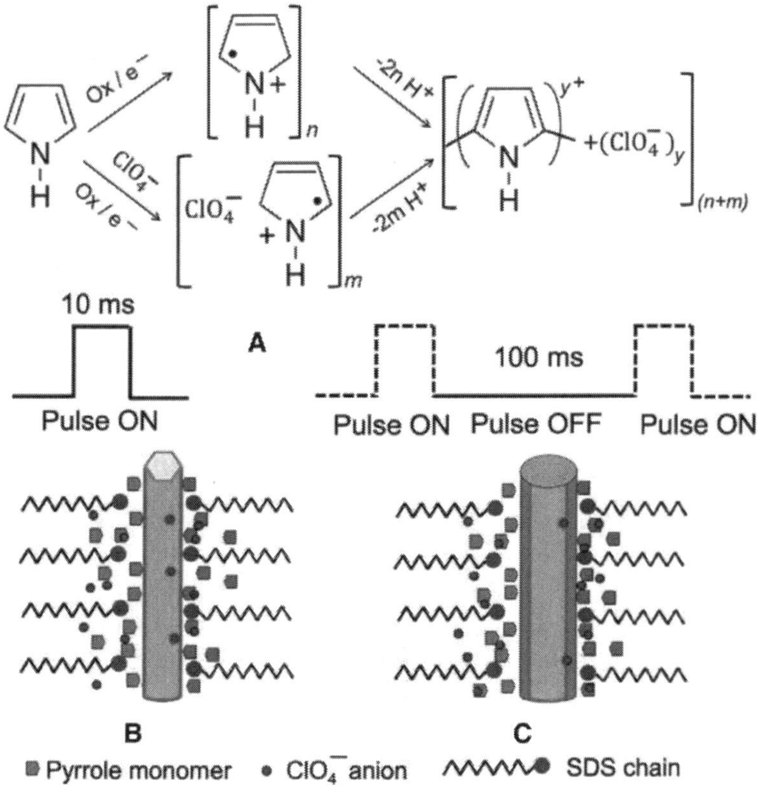

Figure 5 Electropolymerization process of the polypyrrole growth over ZnO nanorods. (A) Electrochemical polymerization of Py monomer and ClO_4 conjugation. **(B)** Model of electropolymerization growth of PPy sheath over ZnO nanorods in the presence of SDS surfactant and **(C)** homogenous growth of PPy sheath over ZnO nanorods after a number of pulsed current cycles.

potentiostatic condition [51, 52]. In the pulsed current electropolymerization process, the polypyrrole growth over ZnO nanorod surface proceeds by concomitant reactions, anodic oxidation of pyrrole monomer, and conjugation reaction with electrolyte (ClO_4^-) anions as shown in Figure 5A. On application of a current pulse of magnitude 4 mA.cm^{-2}, the pyrrole monomer species over the ZnO nanorod surface rapidly oxidize by electron transfer at electrode resulting in the nucleation of significantly large number of $[\mathrm{Py}]_{n+m}^{+\bullet}$ cation radicals. By themselves, these are unstable but stabilize rapidly on interaction with the nearest cation radicals to form short chain oligomers by coupling and bond linkage with the involvement of deprotonation $(-2H^+)_{n+m}$ step [5, 45, 51, 52]. A number of cation radicals $[\mathrm{Py}]_m^{+\bullet}$ at the initiation step are also influenced by strongly interacting electrolyte ClO_4^- anions which result in conjugation of PPy short chain oligomers deposited over ZnO nanorods [53]. The current pulse off time replenishes the Py-monomers at the ZnO nanorods by diffusion in the aqueous medium. The subsequent pulsed current cycle reinitiates the electropolymerization reaction at fresh nucleation sites on ZnO nanorods by a similar process sequence thus providing a uniform coverage.

The preferential nucleation and growth of polypyrrole across the ZnO nanorod length is significantly affected by the lack of access of the pyrrole monomer in deep crevices along the depth of ZnO nanorod array marked by the narrow and not-so-consistent interrod spacing typically varying between 120 to 250 nm. This is further aggravated by aqueous immiscibility of pyrrole monomer which inhibits wetting of ZnO rods which might inhibit formation of uniform polypyrrole sheath. In the present case, the use of SDS anionic surfactant mitigates this by transporting pyrrole monomer to the surface of ZnO nanorods. A possible model of electropolymerization growth of PPy sheath over ZnO nanorods in the presence of SDS surfactant is shown schematically in Figure 5B. The SDS ionizes into Na^+ cation and $CH_3(CH_2)_{11}OSO_3^-$ anion in aqueous medium. The SDS concentration used in this study is less than the critical value 8 mM for the first micelles concentration (CMC-1) hence the SDS molecular chain containing 12 carbon alkyls with sulfate group at the end are in the extended state in the aqueous medium [54, 55]. The dodecyl alkyl molecular chain being hydrophobic orients away from water and this easily attaches on to the ZnO nanorod surface while the hydrophilic OSO_3^- group project outward into aqueous environment. The pyrrole monomers are hydrophobic in character and sparingly soluble in water. A large number of pyrrole monomers are able to preferentially disperse within the hydrophobic region created by attached dodecyl alkyl molecular chain over ZnO nanorod surface [50]. This ensures uninhibited supply of the pyrrole monomer and dopant ClO_4^- anions across the exterior of ZnO nanorods [55] and consequently forming PPy layer over ZnO rods comprising of short-chain doped PPy oligomers by electronation-protonation-conjugation reaction described in Figure 5B. Spatially distributed deposition of PPy oligomers as clusters is evident in the nodule like the microstructure study shown in Figure 2A. The pyrrole monomer availability during current pulsed off time is no longer diffusion-rate limited and efficient incursion of pyrrole results in the increased electropolymerization rates. In the subsequent pulse cycles, the electropolymerization is reinitiated over new ZnO surface sites or over PPy coated surface as shown schematically in Figure 5C resulting in homogenous formation of the PPy sheath over ZnO nanorods after a certain number of current pulsed polymerization cycles.

Cyclic voltammetry study

Figure 6A, B shows a set of CV plots recorded at slow scan rates of 5 and 10 $mV.s^{-1}$ comparing the electrochemical performance of the ZnO nanorod core-PPy sheath electrode with the PPy nanotube structured electrodes obtained by etching ZnO nanorods for 2 and 4 h, respectively. All CV plots are nearly rectangular in shape, symmetrical across the zero current axis, and do not show any oxidation-reduction peaks demonstrating highly capacitive behavior. The current in the PPy nanotube electrodes is significantly higher in comparison to ZnO nanorod-PPy sheath electrode and it further increases for the electrode with an open pore PPy nanotube structure obtained after 4-h etching as shown by the microstructure in Figure 2D. It is apparent that PPy nanotube electrode structure offers improved access to the ions through the tube interior in addition to exterior regions which are accessed equally by all three electrodes. Figure 7A, B shows CV plots measured at scan rates of 5 to 100 $mV.s^{-1}$ for the PPy nanotube electrodes obtained after etching of ZnO nanorods at the core for 2 and 4 h, respectively. The increase in the current with the scan rate indicates the kinetics of the faradic process and the electronic-ionic transport at the PPy nanotube-electrolyte interface. It is easy to observe from Figure 7 that more open PPy nanotube electrodes after 4-h ZnO etch show higher anodic and cathodic current at every scan rate as compared to the 2-h etched electrodes in the same potential window. Although both electrodes showed good charge propagation capabilities, the difference in the current density of the electrodes is attributed to the structural changes due to etching. The CV plots show that though rectangular shape is nearly preserved as the scan rate is increased until 50 $mV.s^{-1}$, a general trend is a progressively narrower and slightly oblique-angled CV plot for scan rate of ≥50 $mV.s^{-1}$. The factors responsible for such a behavior are the contact resistance and the delayed response time of the faradic reactions nonsynchronous with the faster scan which otherwise would have boosted the total capacitance.

The growth in current density of the PPy nanotube electrodes with the increasing scan rate as shown in Figure 7 is reflective of the dissimilarities in terms of the porosity of the nanotube structure and improved performance of the 4-h etched PPy nanotube electrode. The rise in the cathodic peak current density J_{PC} with scan rate, v, follows the Randles-Sevcik equation,

$$J_{PC} = 0.6104 n^{1.5} F c \sqrt{\frac{FD}{RT}} \cdot \sqrt{v} \tag{3}$$

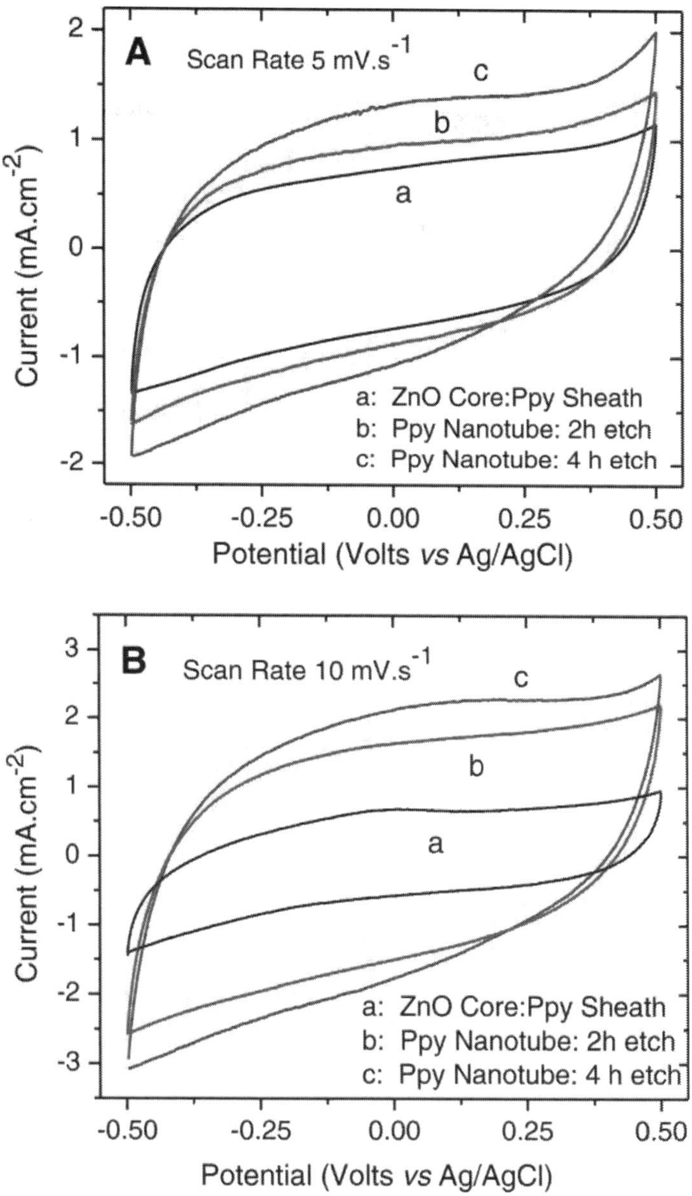

Figure 6 Cyclic voltammetric plots of the electrode with nanostructured ZnO nanorod core-PPy sheath. PPy nanotube after etching away ZnO nanorods for 2 and 4 h measured at scan rates **(A)** 5 mV.s^{-1} and **(B)** 10 mV.s^{-1}.

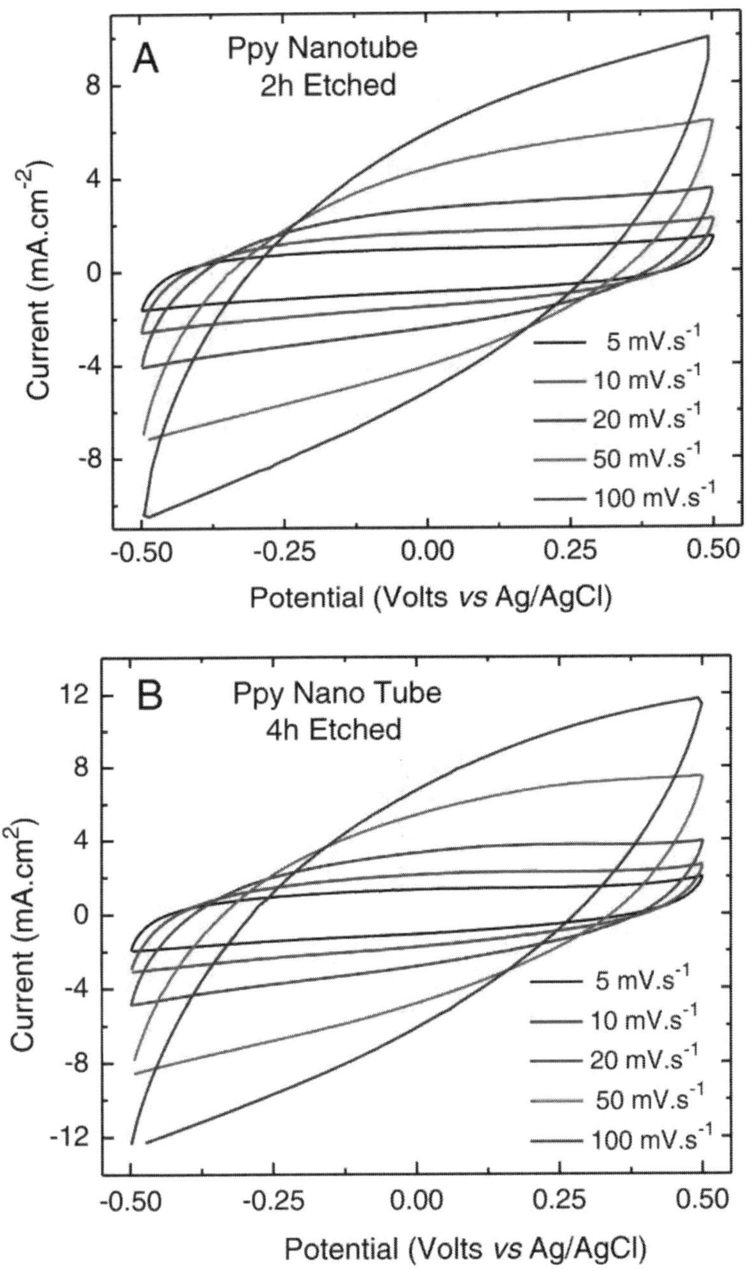

Figure 7 Cyclic voltammetric plots of PPy nanotube electrodes measured at different scan rates. (A) 2-h etch and (B) 4-h etch.

where F is the Faraday number and R is the ideal gas constant. The active specie concentration in electrolyte is denoted by c, and the number of electron-involved reduction processes by n. The parameter D represents the apparent charge transfer coefficient by diffusion. A linear plot of the current *JPCvsv√*

shown in Figure 8 for 2- and 4-h ZnO core etched PPy nanotube electrodes suggests that according to the Randles-Sevcik formulation the charge transport process is diffusive-controlled. Figure 8 shows that compared to the 2-h etched electrode, 4-h etched PPy nanotube electrode has a higher slope which suggests that in this electrode the electrolyte ions are more easily accessible due to the presence of higher diffusivity paths through interconnected nanotubes and therefore have improved ability to store charges. This is consistent with the microstructure study as well as the results of the CV analysis. The measured D values were found to be 7.27×10^{-8} and 1.09×10^{-7} $cm^2.s^{-1}$ for the PPy nanotube structure formed after 2- and 4-h etching, respectively, which is at least an order of magnitude higher than for the PPy films in 2-D porous structure [45]. These data show that homogenous transport dynamics of charge-compensating anions in the electrolyte is generally fast for 3-D PPy nanotubes especially for open interconnected PPy nanotubes formed after 4-h etch.

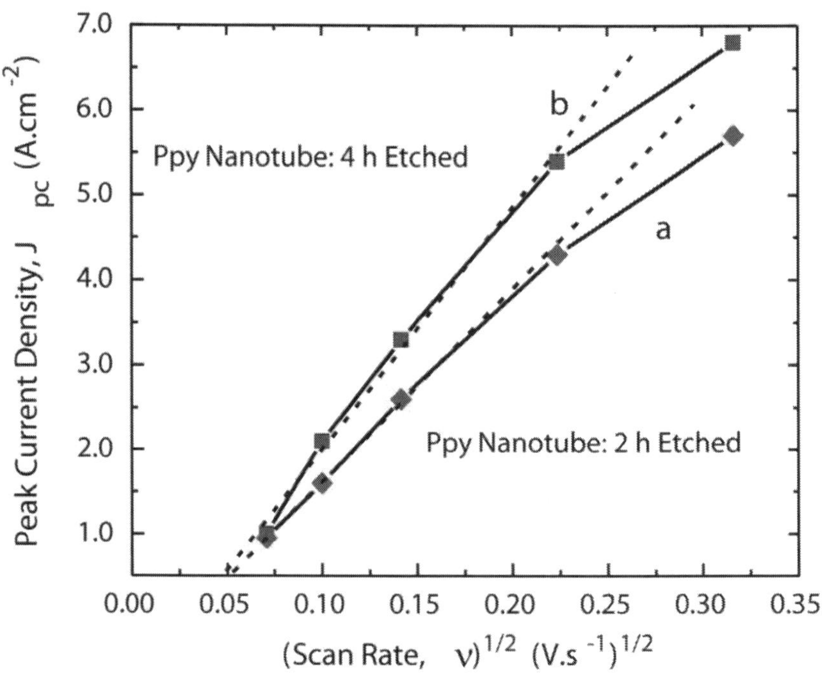

Figure 8 Randles-Sevcik plots of PPy nanotube electrodes after 2- and 4-h etching of ZnO nanorod core.

Specific capacitance C_{SV} calculated from the CV plots using Equation 1 at different scan rates is plotted in Figure 9 for both ZnO nanorod core-PPy sheath and PPy nanotube electrodes represented by 0-, 2-, and 4-h ZnO core etch times. The true faradic specific capacitance due to redox processes measured at low scan rates increases dramatically when the PPy nanostructure transforms from core-sheath to nanotube. Thus, ion diffusion process in PPy nanotube structure is kinetically faster. At higher scan rates ($\geq$50 mV.s^{-1}), the specific capacitance on structure transformation shows moderate increase at best for electrode with open pore PPy nanotube structure obtained after a 4-h ZnO core etch. Limiting kinetics for ion diffusion is the same for PPy sheath and nanotube structures.

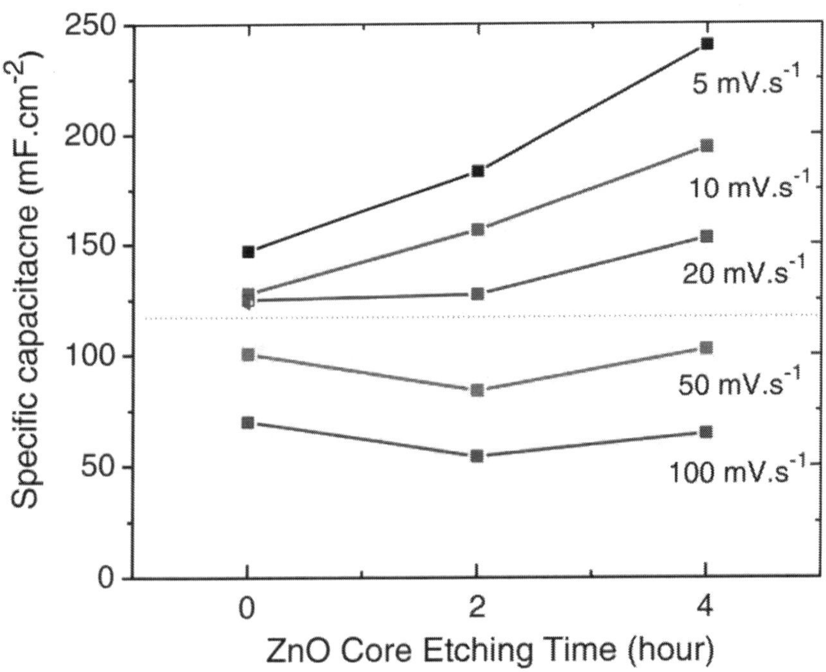

Figure 9 Specific areal capacitance at different scan rates for ZnO nanorod core-PPy sheath PPy and PPy nanotube electrodes.

Impedance spectroscopy

Electrochemical impedance spectroscopy (EIS) technique is extensively used to elucidate the electrical characteristics of the electrode material and its interface with the supporting electrolyte. Frequency response of the real and imaginary impedance of the pseudocapacitive ZnO nanorod core-PPy sheath electrode with 1 M lithium perchlorate electrolyte was studied. Impedance of the electrode is a complex quantity and the extracted data are plotted as real

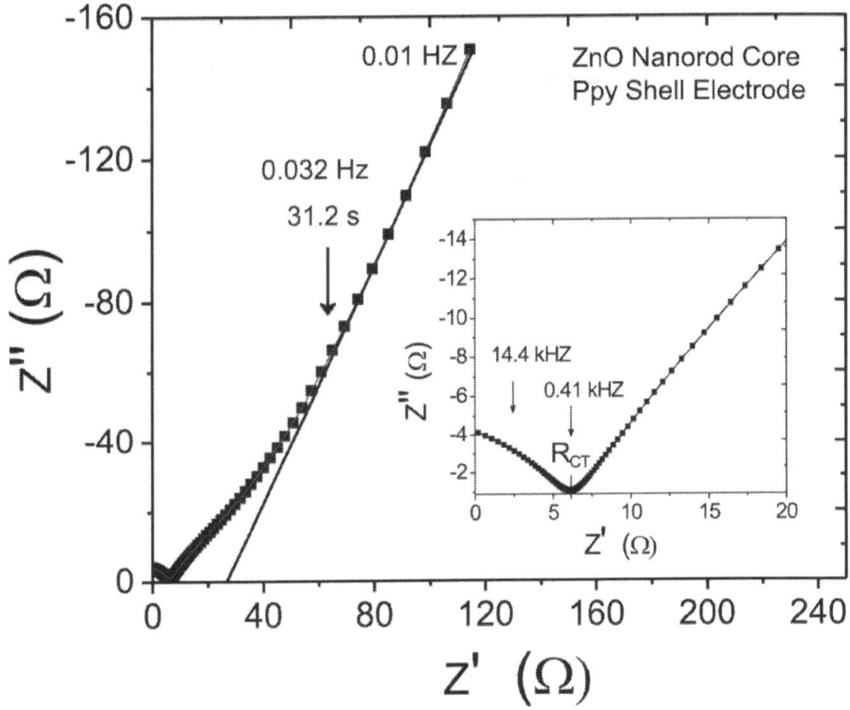

Figure 10 Nyquist plots of actual data and fitted spectrum of ZnO nanorod core-PPy sheath electrode. Inset shows expanded view in the high- and mid-frequency region.

(Z') versus imaginary (Z'') impedance representing the Nyquist plot. Figure 10 shows the Nyquist plot of the as-deposited ZnO nanorod core-PPy sheath electrode in the frequency domain 0.1 MHz to 0.01 Hz and the inset shows expanded view in the high- and mid-frequency region. The capacitive component is reflected in the rapidly increasing imaginary impedance (Z'') at lower frequencies. The high-frequency real impedance (Z') characterizes the bulk electrode and interfacial resistive properties of the electrode-electrolyte system. These parameters calculated from the impedance plots are shown in Table 1. Instead of the characteristic whole semicircle, the high-frequency Nyquist plot degenerated into an arc segment. This suggests that contribution to the bulk electrode-electrolyte resistance is mainly from the ZnO-PPy interface barrier due to polarization effect of the nanostructured electrode and negligible electrolyte resistance. The mid-frequency cutoff of this arc on real impedance Z'-axis gives the charge transfer resistance, R_{CT}, of 6 $\Omega.cm^2$ resulting from the kinetically-controlled electron transfer and anion conjugation reaction in the PPy sheath layer. In progression from the mid- (0.41 kHz) to low-frequency range, a knee frequency of 0.032 Hz is identified indicating the onset of the capacitive impedance. The slow rising impedance in this frequency range is reflective of ion adsorption through the porous structure of the PPy sheath as well as along the length of ZnO nanorods. The capacitive impedance (Z'') shows a shift along more resistive Z' values which is caused by the limitation on the rate of ion migration. Beyond the knee frequency, however, the system response is highly capacitive. The low-frequency areal-capacitance density, C_F, is determined from the Nyquist plot as 107 mF.cm^{-2}.

Table 1 Electrochemical impedance spectroscopy data obtained from actual Nyquist plots

Components	R_s (Ω.cm^2)	R_{ct} (Ω.cm^2)	W (Ω.cm^2)	C_i (mf.cm^{-2})	C_i (f.g^{-1})
ZnO nanorod core-PPy sheath	0	5.8	20.4	107.3	74
Narrow PPy nanotube (2-h etch)	0	8.2	8.4	84.2	58
Open PPy nanotube (4-h etch)	1	7.2	5.4	83	57.2

Figure 11A, B shows the Nyquist plots of the PPy nanotube structure obtained after etching ZnO core for 2 and 4 h, respectively, as described by the SEM study in Figure 2C, D. The major effect of such structural change appears in the shift of the knee frequency to higher frequency values. After 2-h etching with narrow (33 ± 3 nm) PPy nanotube opening and after 4-h etching with open pore interconnected PPy nanotube formation the recorded shifts in knee frequency are 0.16 and 1.07 Hz, respectively, compared to the knee frequency of 0.032 Hz for unetched ZnO nanorod-PPy sheath structured electrode. This shift is significant. Simultaneously, the low-frequency impedance Z'' shows a systematic shift to lower values on the real impedance axis. Considering that knee frequency defines the upper frequency limit of the resistive behavior and a capacitive one at lower than knee frequencies, it is inferred that the PPy nanotube sheath structure is more capacitive in nature. Furthermore, for the unetched ZnO nanorod core-PPy sheath electrodes, the capacitance at knee the frequency is approximately $0.68C_F$ of the overall capacitance C_F. Corresponding values for the 2- and 4-h etched PPy nanotube electrodes are $0.61C_F$ and $0.22C_F$, respectively. These data suggest that over a substantive frequency range the impedance of the PPy nanotube electrode is capacitive in nature. Clearly, the frequency domain of ion diffusion region which resistively contributes to impedance, commonly known as the Warburg resistance, has shrunk in PPy nanotubes after 2-h etching and more significantly in the open interconnected PPy nanotube structure obtained after 4-h etching of ZnO nanorods. It is obvious from the microstructure study of Figure 2C, D and schematic representation in Figure 4 that the interconnected network of 60 to 70 nm diameter open PPy nanotubes facilitates rapid and more pervasive ingression of anions from electrolyte as frequency decreases below the knee

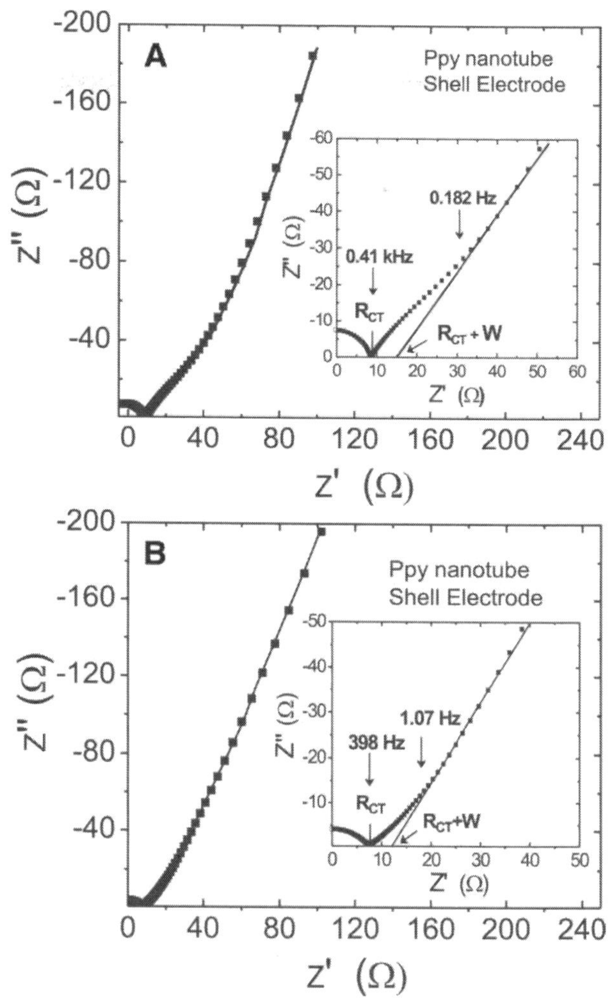

Figure 11 Nyquist plots of actual data and fitted spectrum of PPy nanotube electrodes obtained after etching ZnO core. (A) **2 h and** (B) **4 h.**

frequency. As frequency decreases, electrolyte ions by diffusion are accessible to more and deeper porous surface of the PPy nanotube arrays. The frequency response of the impedance is modeled in terms of complex capacitance $C(\omega) = C'(\omega) - jC''(\omega)$ to describe the capacitance behavior of the electrodes [56]. Here, $C'(\omega)$ is the real part of capacitance representing the energy storage component and $C''(\omega)$ the imaginary part represents the resistive losses in the storage process. The real capacitance is computed according to equation $C'(\omega) = [-Z''(\omega)]/[\omega|Z(\omega)|^2]$. Figure 12 shows variation of C'/C_0 as a function of frequency, where C_0 is dc capacitance [57]. As the frequency decreases, C' sharply increases below and above 1 Hz, the capacitance is practically nonexistent. Figure 12 also shows phase angle variation with frequency. The low-frequency phase angle shows a plateau at -65° for PPy nanotube sheath electrode after 4-h etching which indicates a capacitor-like behavior though not yet an ideal one for which phase angle should be closer to -90°. Compared to the nonplateau behavior and low phase angle of -40° observed in the unetched ZnO nanorod core-PPy sheath electrode, the PPy nanotube electrode shows considerably improved capacitor behavior.

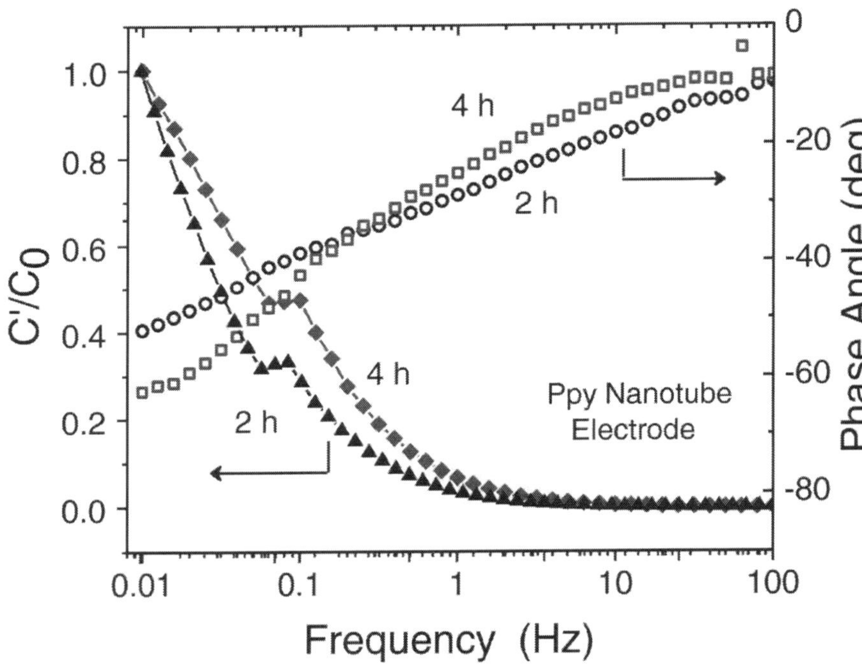

Figure 12 Frequency dependence of areal-specific capacitance to dc capacitance and phase angle variation for PPy nanotube electrodes.

The measured charge transfer resistance, R_{CT}, is 8.2 and 7.2 Ω cm^2, respectively, for 2- and 4-h etched PPy nanotube structured electrodes, which is not much different from that of the unetched ZnO nanorod core-PPy sheath structured electrode. It is obvious that extent of anion conjugation reaction in the PPy nanotube sheath in response to the electron transfer action is not much affected as the ZnO core is etched away. A more significant effect of the PPy nanotube sheath is seen in the Warburg impedance values. The intercept of extrapolation of the low-frequency impedance on the *x*-axis gives resistance $R_{CT} + W$, where W is the Warburg impedance. As shown in Table 1, W equals 20.2 Ω.cm^2 for unetched ZnO nanorods core-PPy sheath electrode and decreases to 8.4 and 5.4 Ω.cm^2 for the PPy nanotube structure realized after 2- and 4-h etching, respectively.

The impedance parameters of the complex ZnO nanorod core-PPy sheath electrode system were analyzed by equivalent circuit modeling. Nyquist plots are simulated using the equivalent circuit shown in Figure 13 and the component parameters were derived that provide closest fit at each frequency point [58]. Electrical equivalence of the physical electrochemical entities is marked in Figure 13. The equivalent circuit model includes solution resistance R_S, charge transfer resistance R_{CT} representing the electrode kinetics, and Warburg element CPE_W representing the resistance encountered in diffusion and access of ions within nanoporous electrode structure. The inclusion of the constant phase element CPE_{dl} instead of the conventional purely capacitive element C_{dl} is to account for the dispersive behavior of the capacitance arising from the charge accumulation layer at the ZnO nanorods exposed to the electrolyte through pores in the PPy sheath and nanostructure of the electrode. Similarly, CPE_{nr} is the capacitive element which characterizes the pseudocapacitance property of the nanotubular PPy-anion conjugation. The nanostructure resistance, R_{nr}, is representative of the electron transport resistance due to narrow (approximately 60 nm diameter) vertically long (approximately 2.2 μm) ZnO nanorods and C_{nr} its electrochemical capacitance [59]. The continuous lines in the Nyquist plots in Figures 10 and 11 are the results of the fitting based on this model. Excellent fit is observed over the entire frequency range. Various electrical resistance and capacitive parameters estimated by fitting of Nyquist plots are summarized in Table 2.

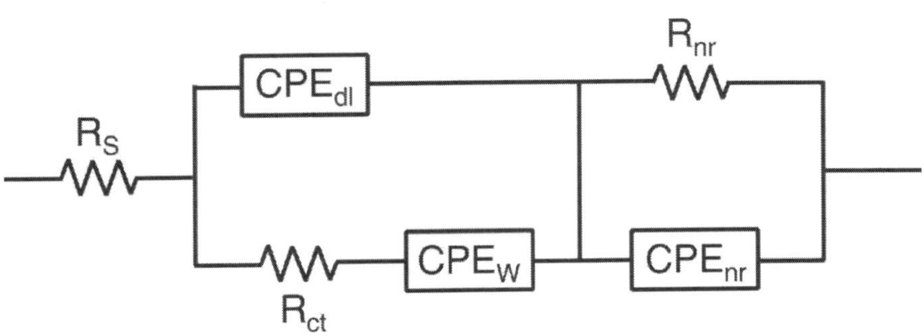

Figure 13 Equivalent electric circuit model used for simulation of Nyquist plots.

Table 2 Characteristic resistance and capacitive parameters estimated by fitting of Nyquist plots.

Components	CPE_{dl} (mMho, p)	R_{ct} (Ω)	CPE_w (mMho, p)	R_{nr} (Ω)	CPE_{nr} (mMho, p)
ZnO nanorod core-PPy sheath	$Q – 0.025$ $p – 0.55$	21.24	$Q – 0.03$ $p – 0.61$	6	$Q – 0.012$ $p – 0.75$
Narrow PPy nanotube (2-h etch)	$Q – 0.0006$ $p – 0.87$	18	$Q – 0.036$ $p – 0.74$	28	$Q – 0.065$ $p – 0.44$
Open PPy nanotube (4-h etch)	$Q – 0.04$ $p – 0.61$	16	$Q – 0.04$ $p – 0.76$	20	$Q – 0.389$ $p – 0.42$

The constant phase element (CPE) instead of the capacitor in the equivalent circuit above is justified in order to more appropriately account for the heterogeneities including the surface roughness, porosity, and variation in the PPy thickness arising from the nanostructured nature of the ZnO-PPy electrode. The long, vertical, and dispersed 3-D ZnO nanorod core-PPy sheath (nanotube) nanostructure has a diverse aspect ratio relative to a flat 2-D electrode structure and therefore differently impacts the ion diffusion kinetics. This gives rise to the distributed time constants simulating the capacitance dispersion which is better represented by the RC network comprising of nanostructure resistance, R_{nr}, and the constant phase element, CPE_{nr}[60]. The CPE_{nr} impedance is given as, [61].

$$Z''_{\mathrm{CPE_{nr}}} = \frac{1}{(j\omega)^p \cdot Q} \tag{4}$$

where exponent p represents dispersive nature of time constant, since with $p = 1$, the impedance Z'' is purely capacitive characterized by a single time constant and the parameter Q is equivalent to a capacitance, while for $p < 1$ parameter Q is basically a CPE with units $Mho.cm^{-2}$. Figure 14 shows the log-log plot of measured Z'' versus frequency data for ZnO nanorod core-PPy sheath and PPy nanotube electrodes which is a straight line according to above Equation 4, and the slope corresponds to exponent, p. The Z'' calculated from the simulated exponent p and CPE Q values according to above equation and measured Z'' values are compared in Table 3. The low-frequency measured impedance, Z'', shows a deviation by a constant factor from the simulated CPE_{nr} impedance which is attributed to C_{nr} which has weak frequency dispersion but contributes to the measured Z'' values. The exponent p in the simulation of the impedance due to the Warburg element has the values $0.5 < p < 1$ (Table 2). A $p = 0.5$ usually defines the CPE_W due to Warburg impedance. The higher simulated p values are interpreted as diffusion-based resistance signifying the pseudocapacitive nature of electrodes.

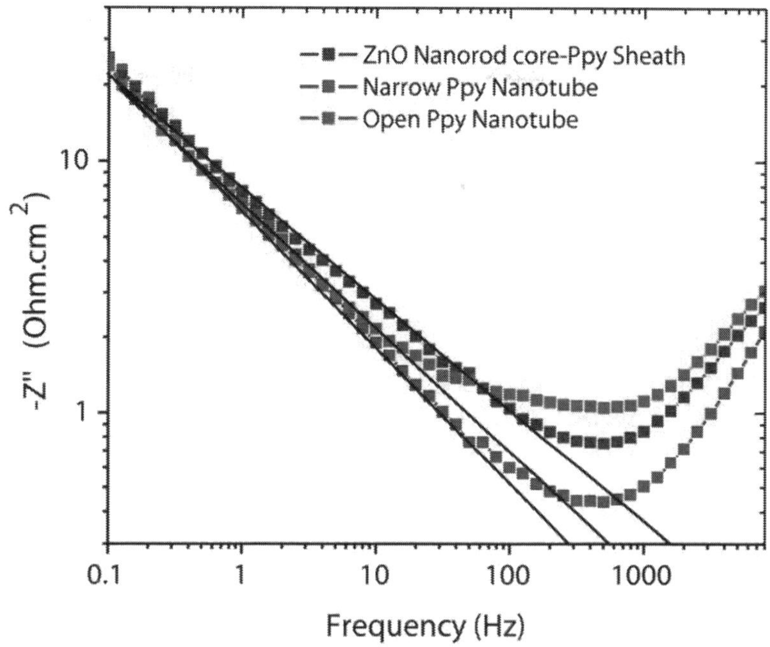

Figure 14 The log-log Bode plot of measured data for ZnO nanorod core-PPy sheath and PPy-nanotube electrodes at various frequencies.

Table 3 The **Z″** parameter calculated from the simulated exponent **p** and CPE **Q** values and the measured **Z″** values.

Nanostructure electrode	Simulated $Z''_{CPE_{or}}$ mhos.cm^{-2}s^{-p}	Measured $Z''_{CPE_{or}}$ mhos.cm^{-2}s^{-p}	Percentage deviation
ZnO nanorod core-PPy sheath	11.8	7.7	34.7
Narrow PPy nanotube (2-h etch)	11.2	7.15	36.1
Open PPy nanotube (4-h etch)	9.59	6.50	32.2

Charge-discharge curves and stability analysis

Galvanostatic charge-discharge performance of the ZnO nanorod core-PPy sheath and PPy nanotube electrodes was studied at various current densities from 1 to 3 mA.cm^{-2} in the voltage range 0.05 to 0.5 V. Figure 15A shows charge-discharge curves measured at 1 mA.cm^{-2} for the PPy nanotube electrode obtained by a 2- and 4-h etch and its comparison with that of the ZnO nanorod core-PPy sheath electrode. The discharge curves are nearly linear for all the three electrode structures indicating highly capacitive character. The areal-capacitance density C_{sd} of the electrodes was calculated from the charge-discharge curves at 1 mA.cm^{-2} using Equation 2 and the results are presented in

Table 4. The electrode with PPy open nanotube structure is higher than the electrode with ZnO nanorod core-PPy sheath structure. This suggests that electrolyte ions are able to access through nanotubes and can intercalate better with PPy taking advantage of the exposed nanotube surface. A small IR drop at the start of the discharge cycle is noticed in each of these electrodes which is due to equivalent series resistance (ESR) arising from contacts and internal electrode cell resistance. Typical ESR values are in the range 25 to 40 $\Omega.cm^2$ as shown in Table 4. The internal resistance contribution to the ESR originates from the core and the thin ZnO seed layer at the graphite substrate. In the charge (doping) cycle extraction and in the dedoping (discharge) cycle, the ingress of electrons from and into the PPy tubular shell has a percolation path through ZnO rods which offers a finite resistance. This is evident from the significantly lower ESR values in the electrodes with the partial dissolution of ZnO core with narrow opened PPy nanotubes obtained after a 2-h etch. A slight increase ESR is observed in the electrodes with open PPy nanotube structure. The contact between PPy sheath over ZnO nanorods and the others in the vicinity is minimal at best as the sheath thickness is on the average less than the inter-ZnO nanorod spacing (see Figures 2, 3 and 4). After complete dissolution of ZnO, the finite contact resistance between the freestanding PPy nanotube sheaths is responsible for increase in ESR. The effect of charge-discharge current density on the charge-discharge characteristics for each of these electrodes in ZnO nanorod core-PPy sheath PPy nanotube structures is shown in Figure 15B, C, D which follows a similar trend as discussed in the context of Figure 15A. The specific capacitances of these electrodes were calculated at different constant current density and the results are plotted in Figure 16 as a function of discharge current density. In the case of PPy nanotube electrodes, a decrease in the specific capacitance with increasing discharge current is observed. This suggests that the redox process is kinetically dependent on the ionic diffusion at the PPy nanotube-electrolyte interface even though the nanotubes have an unabated access to the ions as evident from the increased specific capacitance of the electrode with open PPy nanotube structure over the one having narrow PPy nanotube structure. The nearly constant specific capacitance of the ZnO nanorod core-PPy sheath electrode with increasing discharge current density is suggestive of faster redox kinetics at the interface. These observations suggest that the redox process in the PPy nanotube electrodes is due to limitation on electron transport rather than the diffusive access of electrolyte dopant ions to the PPy in the nanotube structure. The electron transport is facilitated through ZnO nanorods in close contact with graphite substrates. In the case of PPy nanotubes, electron transport can only take place through the PPy nanotube along its length. Since anion conjugation (doping) is in response to the electron extraction in spite of unimpeded access to electrolyte anions, the doping process is limited by electron transport. The reduction in the specific capacitance in PPy nanotubes at higher charge current and the increase in specific capacitance of 3-D ZnO nanorod PPy sheath structure electrode with the increase in charging current as observed in Figure 16 are explicable on this basis.

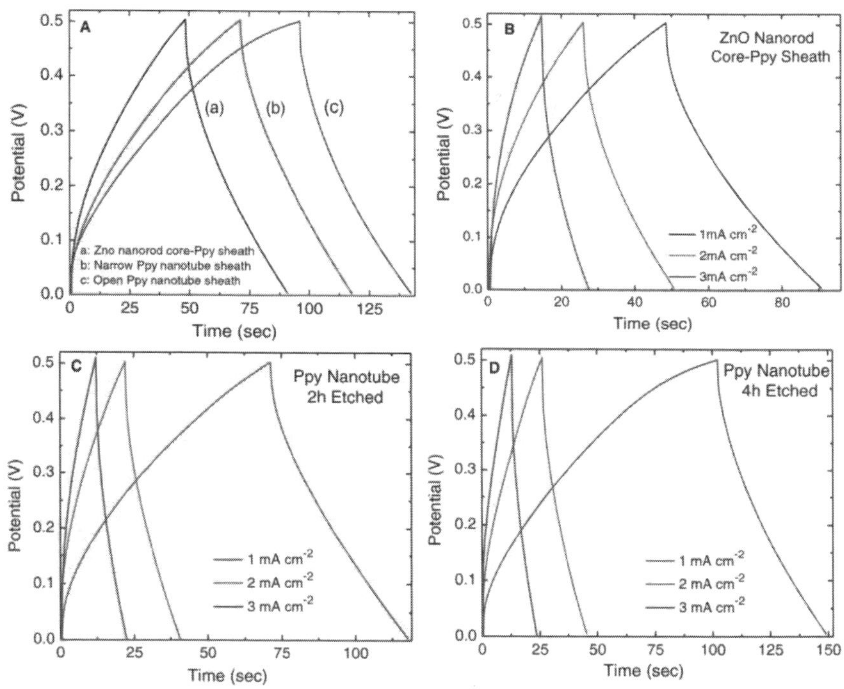

Figure 15 Charge-discharge characteristics. (A) ZnO nanorod core PPy sheath electrode and PPy nanotube electrodes after 2-h and 4-h etch measured at a constant current density of 1 $mA.cm^{-2}$. Charge-discharge characteristics measured at different current densities for **(B)** ZnO nanorod core-PPy sheath, **(C)** PPy nanotube 2-h etch, and **(D)** PPy nanotube 4-h etch.

Table 4 Specific areal capacitance and ESR of nanostructured electrodes calculated from charge-discharge curves measured at 1 $mA.cm^{-2}$

Nanostructure electrode	C_{sd} ($mF.cm^{-2}$)	ESR ($\Omega.cm^2$)
ZnO nanorod core-PPy sheath	131.22	40.5
Narrow PPy nanotube (2-h etch)	132.28	25.08
Open PPy nanotube (4-h etch)	141.09	32.09

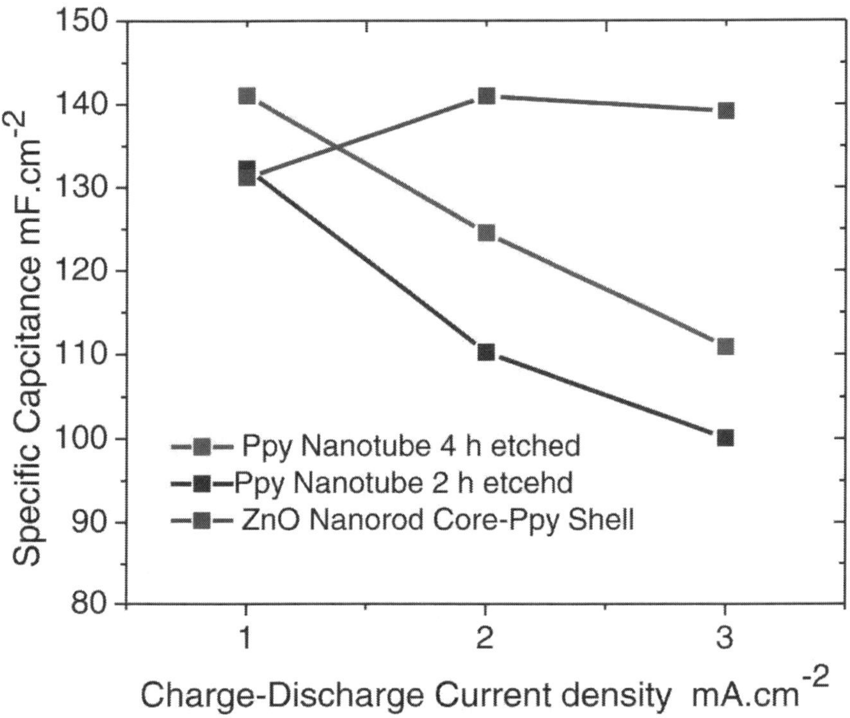

Figure 16 The specific capacitances of the ZnO nanostructured electrodes plotted as a function of charge-discharge current density.

Cycling test

The cycling stability of the open PPy nanotube electrode was investigated at a constant charge-discharge current density of 1 mA.cm^{-2} for a continuous 5,000 cycles. Figure 17 shows the effect on the discharge capacitance density as a function of the number of charge-discharge cycle. The overall change in the discharge capacitance is only <12% indicative of highly stable redox performance and electrochemical stability of the PPy nanotube electrode. This stability arises from unimpeded access of the electrolyte ions through diffusive transport across to a large fraction of the PPy polymer surface due to the 3-D nanotube structure in the redox process. Furthermore, the PPy nanotube electrodes do not show physical or chemical degradation during cycling. This is borne out from the ESR data, which remains on the average nearly constant during cycling tests for 5,000 cycles.

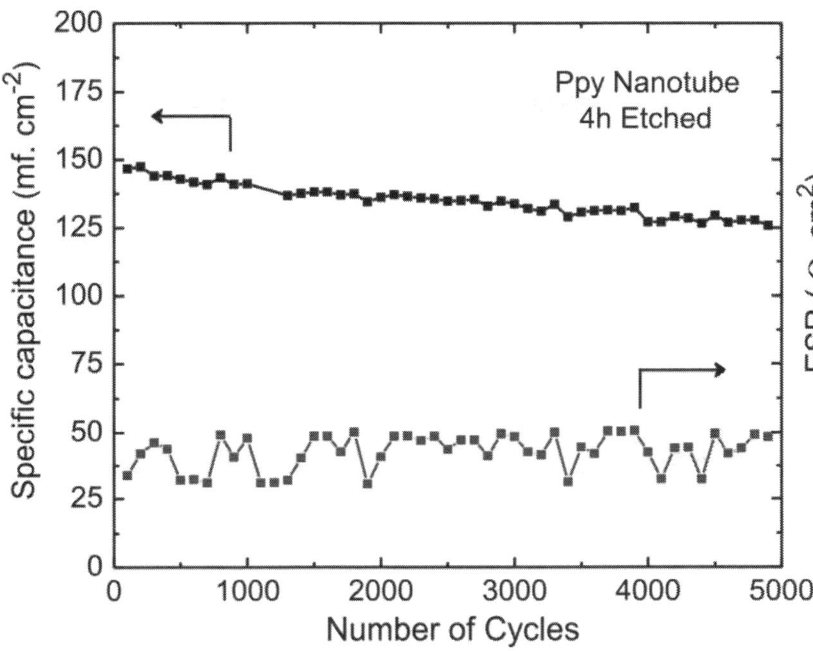

Figure 17 Long-term charge-discharge cycle tests for PPy nanotube 4-h etched electrode showing discharge capacitance density and ESR variation.

CONCLUSIONS

Electrodes in the three-dimensional nanoscale architecture studied in this work in the form of vertically aligned ZnO nanorod PPy sheath and PPy nanotube show considerable potential for high energy-density storage in a supercapacitor device. These nanostructures are formed by depositing a sheath of PPy over vertical ZnO nanorod arrays by controlled pulsed current electropolymerization and by selective etching of the ZnO nanorod core. Based on the cyclic voltammetry data, electrode with open interconnected PPy nanotube array structure shows high areal-specific capacitance of approximately 240 mF.cm^{-2} attributed to realization of enhanced access to electrolyte ions. The observed scan rate dependence of the current has been interpreted as delayed response time of faradic reaction nonsynchronous with faster scan rate, which could possibly have boosted capacitance density further. Slow redox processes are shown to be due to limitation of electron transfer across the length of vertical PPy nanotube arrays rather than the diffusive transport of electrolyte ions. Managing this limitation could possibly enhance the specific capacitance and thus energy storage ability further.

ACKNOWLEDGEMENTS

This work was supported by the National Science Foundation (NSF) grant in aid project 1318202, the Partnership for Innovation in Electrochemical Energy Storage, and the Office of Naval Research (ONR) under contract N00014-11-1-0658 which are gratefully acknowledged.

REFERENCES

1. Zhang LL, Zhao XS: Carbon-based materials as supercapacitor electrodes. Chem Soc Rev 2009, 38: 2520–2531.
2. Conway BE: Electrochemical Supercapacitors: Scientific Fundamentals and Technological Applications. New York: Springer; 1999.
3. Snook GA, Kao P, Best AS: Conducting-polymer-based supercapacitor devices and electrodes. J Power Sources 2011, 196: 1–12.
4. Wang G, Zhang L, Zhang J: A review of electrode materials for electrochemical supercapacitors. Chem Soc Rev 2012, 41: 797–828.
5. Pandey GP, Rastogi AC: Synthesis and characterization of pulsed polymerized poly(3,4-ethylenedioxythiophene) electrodes for high-performance electrochemical capacitors. Electrochimica Acta 2013, 87: 158–168.
6. Bae J, Song MK, Park YJ, Kim JM, Liu M, Wang ZL: Fiber supercapacitors made of nanowire-fiber hybrid structures for wearable/flexible energy storage. Angew Chem Int Ed 2011, 50: 1683–1687.
7. Tao J, Liu N, Ma W, Ding L, Li L, Su J, Gao Y: Solid-state high performance flexible supercapacitors based on polypyrrole-MnO_2-carbon fiber hybrid structure. Sci Rep 2013, 3
8. Wang K, Wu H, Meng Y, Wei Z: Conducting polymer nanowire arrays for high performance supercapacitors. Small Weinh Bergstr Ger 2014, 10: 14–31.
9. Li G, Peng H, Wang Y, Qin Y, Cui Z, Zhang Z: Synthesis of polyaniline nanobelts. Macromol Rapid Commun 2004, 25: 1611–1614.
10. Simon P, Gogotsi Y: Materials for electrochemical capacitors. Nat Mater 2008, 7: 845–854.
11. Sidhu NK, Rastogi AC: Nanoscale blended MnO_2 nanoparticles in electro-polymerized polypyrrole conducting polymer for energy storage in supercapacitors. MRS Online ProcLibr 2013, 1552: 11–16.

12. Sharma RK, Rastogi AC: Manganese oxide embedded polypyrrole nanocomposites for electrochemical supercapacitor. Electrochimica Acta 2008, 53: 7690–7695.

13. Pintu Sen AD: Electrochemical performances of poly(3,4-ethylenedioxythiophene)–$NiFe_2O_4$ nanocomposite as electrode for supercapacitor. Electrochimica Acta 2010, 55: 4677–4684.

14. Lee SW, Kim J, Chen S, Hammond PT, Shao-Horn Y: Carbon nanotube/manganese oxide ultrathin film electrodes for electrochemical capacitors. ACS Nano 2010, 4: 3889–3896.

15. Wang Y, Guo CX, Liu J, Chen T, Yang H, Li CM: CeO_2 nanoparticles/graphene nanocomposite-based high performance supercapacitor. Dalton Trans 2011, 40: 6388–6391.

16. Cheng Q, Tang J, Ma J, Zhang H, Shinya N, Qin L-C: Graphene and nanostructured MnO_2 composite electrodes for supercapacitors. Carbon 2011, 49: 2917–2925.

17. Yang D: Application of nanocomposites for supercapacitors: characteristics and properties. In Nanocomposites - New Trends Dev. Edited by: Ebrahimi F. Rijeka: InTech; 2012:299–328.

18. Wu N-L: Nanocrystalline oxide supercapacitors. Mater Chem Phys 2002, 75: 6–11.

19. An J, Liu J, Ma Y, Li R, Li M, Yu M, Li S: Fabrication of graphene/polypyrrole nanotube/MnO_2 nanotube composite and its supercapacitor application. Eur Phys J Appl Phys 2012, 58.

20. Zhu J, Shi W, Xiao N, Rui X, Tan H, Lu X, Hng HH, Ma J, Yan Q: Oxidation-etching preparation of MnO_2 tubular nanostructures for high-performance supercapacitors. ACS Appl Mater Interfaces 2012, 4: 2769–2774.

21. Liu J, Jiang J, Cheng C, Li H, Zhang J, Gong H, Fan HJ: Co_3O_4 Nanowire@MnO_2 ultrathin nanosheet core/shell arrays: a new class of high-performance pseudocapacitive materials. Adv Mater 2011, 23: 2076–2081.

22. Xiao X, Ding T, Yuan L, Shen Y, Zhong Q, Zhang X, Cao Y, Hu B, Zhai T, Gong L, Chen J, Tong Y, Zhou J, Wang ZL: WO_{3-x}/MoO_{3-x}core/shell nanowires on carbon fabric as an anode for all solid state asymmetric supercapacitors. Adv Energy Mater 2012, 2: 1328–1332.

23. Kim J-H, Zhu K, Yan Y, Perkins CL, Frank AJ: Microstructure and pseudocapacitive properties of electrodes constructed of oriented NiO-TiO_2 nanotube arrays. Nano Lett 2010, 10: 4099–4104.

24. Dong X, Cao Y, Wang J, Chan-Park MB, Wang L, Huang W, Chen P: Hybrid structure of zinc oxide nanorods and three dimensional graphene foam for supercapacitor and electrochemical sensor applications. RSC Adv 2012, 2: 4364–4369.

25. Xiao R, Cho SI, Liu R, Lee SB: Controlled electrochemical synthesis of conductive polymer nanotube structures. J Am Chem Soc 2007, 129: 4483–4489.

26. Liu J, An J, Ma Y, Li M, Ma R: Synthesis of a graphene-polypyrrole nanotube composite and its application in supercapacitor electrode. J Electrochem Soc 2012, 159

27. Pan L, Qiu H, Dou C, Li Y, Pu L, Xu J, Shi Y: Conducting polymer nanostructures: template synthesis and applications in energy storage. Int J Mol Sci 2010, 11: 2636–2657.

28. Pan LJ, Pu L, Shi Y, Song SY, Xu Z, Zhang R, Zheng YD: Synthesis of polyaniline nanotubes with a reactive template of manganese oxide. Adv Mater 2007, 19: 461–464.

29. Salari M, Aboutalebi SH, Konstantinov K, Liu HK: A highly ordered titania nanotube array as a supercapacitor electrode. Phys Chem Chem Phys 2011, 13: 5038–5041.

30. Yiwen Tang LL: Electrodeposition of ZnO nanotube arrays on TCO glass substrates. Electrochem Commun 2007, 9: 289–292.

31. Liu P, Zhang H, Liu H, Wang Y, Yao X, Zhu G, Zhang S, Zhao H: A facile vapor-phase hydrothermal method for direct growth of titanate nanotubes on a titanium substrate via a distinctive nanosheet roll-up mechanism. J Am Chem Soc 2011, 133: 19032–19035.

32. Vayssieres L: Growth of arrayed nanorods and nanowires of ZnO from aqueous solutions. Adv Mater 2003, 15: 464–466.

33. Wang Z-L, He X-J, Ye S-H, Tong Y-X, Li G-R: Design of polypyrrole/polyaniline double-walled nanotube arrays for electrochemical energy storage. ACS Appl Mater Interfaces 2014, 6: 642–647.

34. Sidhu NK, Thankalekshmi RR, Rastogi AC: Solution processed TiO_2 nanotubular core with polypyrrole conducting polymer shell structures for supercapacitor energy storage devices. MRS Online Proc Libr 2013, 1547: 69–74.

35. Kim MS, Park JH: Polypyrrole/titanium oxide nanotube arrays composites as an active material for supercapacitors. J Nanosci Nanotechnol 2011, 11: 4522–4526.

36. Wang Z-L, Guo R, Ding L-X, Tong Y-X, Li G-R: Controllable template-assisted electrodeposition of single- and multi-walled nanotube arrays for electrochemical energy storage. Sci Rep 2013.,

37. Yang Y, Kim D, Yang M, Schmuki P: Vertically aligned mixed V_2O_5–TiO_2 nanotube arrays for supercapacitor applications. Chem Commun 2011, 47: 7746–7748.

38. Cho SI, Lee SB: Fast electrochemistry of conductive polymer nanotubes: synthesis, mechanism, and application. Acc Chem Res 2008, 41: 699–707.

39. Zhao Z, Lei W, Zhang X, Wang B, Jiang H: ZnO-based amperometric enzyme biosensors. Sensors 2010, 10: 1216–1231.

40. Choi Y-S, Kang J-W, Hwang D-K, Park S-J: Recent advances in ZnO-based light emitting diodes. IEEE Trans Electron Devices 2010, 57: 26–41.

41. Thankalekshmi RR, Dixit S, Rastogi AC: Doping sensitive optical scattering in zinc oxide nanostructured films for solar cells. Adv Mater Lett 2013, 4: 9.

42. Pearton SJ, Norton DP, Heo YW, Tien LC, Ivill MP, Li Y, Kang BS, Ren F, Kelly J, Hebard AF: ZnO spintronics and nanowire devices. J Electron Mater 2006, 35: 862–868.

43. Thankalekshmi RR, Dixit S, Rastogi AC, Samanta K, Katiyar RS: Closed-space flux sublimation growth and properties of (Cu-Mn)-doped ZnO thin films in nanoneedle-like morphologies. Integr Ferroelectr 2011, 125: 130.

44. Wang ZL: Zinc oxide nanostructures: growth, properties and applications. J Phys Condens Matter 2004, 16: R829.

45. Sharma RK, Rastogi AC, Desu SB: Pulse polymerized polypyrrole electrodes for high energy density electrochemical supercapacitor. Electrochem Commun 2008, 10: 268–272.

46. Singh A, Joshi A, Samanta S, Debnath AK, Aswal DK, Gupta SK, Yakhmi JV: Charge transport in polypyrrole: ZnO-nanowires composite films. Appl Phys Lett 2009, 95: 2106.

47. Dong J-J, Zhen C-Y, Hao H-Y, Xing J, Zhang Z-L, Zheng Z-Y, Zhang X-W: Controllable synthesis of ZnO nanostructures on the Si substrate by a hydrothermal route. Nanoscale Res Lett 2013, 8: 378.

48. Gurunathan K, Murugan AV, Marimuthu R, Mulik UP, Amalnerkar DP: Electrochemically synthesised conducting polymeric materials for

applications towards technology in electronics, optoelectronics and energy storage devices. Mater Chem Phys 1999, 61: 173–191.

49. Zhou M, Heinze J: Electropolymerization of pyrrole and electrochemical study of polypyrrole: 1. Evidence for structural diversity of polypyrrole. J Electrochem Acta 1999, 44: 1733–1748.

50. Dai T, Yang X, Lu Y: Controlled growth of polypyrrole nanotubule/wire in the presence of a cationic surfactant. Nanotechnology 2006, 17

51. Sadki S, Schottland P, Brodie N, Sabouraud G: The mechanisms of pyrrole electropolymerization. Chem Soc Rev 2000, 29: 283–293.

52. Genies EM, Bidan G, Diaz AF: Spectroelectrochemical study of polypyrrole films. J Electroanal Chem Interfacial Electrochem 1983, 149: 101–113.

53. Qiu Y-J, Reynolds JR: Electrochemically initiated chain polymerization of pyrrole in aqueous media. J Polym Sci Part Polym Chem 1992, 30: 1315–1325.

54. Hazarika J, Kumar A: Controllable synthesis and characterization of polypyrrole nanoparticles in sodium dodecylsulphate (SDS) micellar solutions. Synth Met 2013, 175: 155–162.

55. Naoi K, Oura Y, Maeda M, Nakamura S: Electrochemistry of surfactant�doped polypyrrole film(I): formation of columnar structure by electropolymerization. J Electrochem Soc 1995, 142: 417–422.

56. Taberna PL, Simon P, Fauvarque JF: Electrochemical characteristics and impedance spectroscopy studies of carbon-carbon supercapacitors. J Electrochem Soc 2003, 150

57. Taberna PL, Portet C, Simon P: Electrode surface treatment and electrochemical impedance spectroscopy study on carbon/carbon supercapacitors. Appl Phys A 2006, 82: 639–646.

58. Hrdy R, Kynclova H, Drbohlavova J, Svatos V, Chomoucka J, Prasek J, Businova P, Pekarek J, Trnkova L, Kizek R: Electrochemical impedance spectroscopy behaviour of guanine on nanostructured planar electrode. J Electrochem Sci 2013, 8: 4384–4396.

59. Martinson ABF, Góes MS, Fabregat-Santiago F, Bisquert J, Pellin MJ, Hupp JT: Electron transport in dye-sensitized solar cells based on ZnO nanotubes: evidence for highly efficient charge collection and exceptionally rapid dynamics. J Phys Chem A 2009, 113: 4015–4021.

60. Jorcin J-B, Orazem ME, Pébère N, Tribollet B: CPE analysis by local electrochemical impedance spectroscopy. Electrochimica Acta 2006, 51: 1473–1479.

61. Hirschorn B, Orazem ME, Tribollet B, Vivier V, Frateur I, Musiani M: Constant-phase-element behavior caused by resistivity distributions in films I. Theory. J Electrochem Soc 2010, 157

Chapter 9

A Hydrogen Peroxide Sensor Prepared by Electropolymerization of Pyrrole Based on Screen-Printed Carbon Paste Electrodes

Guang Li [1,*], You Wang [2] and Hui Xu [1]

[1] *National Laboratory of Industrial Control Technology, Zhejiang University, Hangzhou 310027, P.R. China*

[2] *Department of Biomedical Engineering, Zhejiang University, Hangzhou 310027, P.R. China*

ABSTRACT

A disposable amperometric biosensor for commercial use to detect hydrogen peroxide has been developed. The sensor is based on screen-printed carbon paste electrodes modified by electropolymerization of pyrrole with horseradish peroxidase (HRP) entrapped. The facture techniques of fabricating the enzyme electrodes are suitable for mass production and quality control. The biosensor shows a linear amperometric response to H_2O_2 from 0.1 to 2.0 mM, with a sensitivity of 33.24 μA mM^{-1} cm^{-2}. Different operational parameters of electropolymerization are evaluated and optimized.

Keywords: Electropolymerization; Biosensor; Polypyrrole; Horseradish Peroxidase; Screen-Printed Electrodes

1. INTRODUCTION

The determination of hydrogen peroxide is of considerable interest, because hydrogen peroxide is not only an important analyte in food, pharmaceutical, clinical, industrial and environmental analyses but also playing a key role as the product of the enzymatic reaction in coupled enzyme systems [1]. Several analytical techniques have been employed for this determination, such as titrimetry [2], spectrometry [3], chemiluminescene [4, 5], but these techniques suffer from interferences, long analysis time and use of expensive reagents. Electroanalytical methods [6-11] have also been found suitable since they achieve low detection limits and rapid response time. Coupled with enzymatic reactions, it is promising for fabrication of simple and low-cost enzyme sensors. Until now, horseradish peroxidase (HRP) has become a commonly used enzyme to construction hydrogen peroxide biosensors.

It is well known that direct electron transfer from the reduced enzyme to a distant electrode is negligible. To improve it, redox enzyme catalysis can be mediated by a variety of organic and inorganic species which act efficiently in competition with the natural acceptors (or donors) of electrons [12]. The most widely used mediators are ferrocene derivatives [9, 13], although other types are employed [14].

The technique to immobilize the enzyme is one of the key issues in developing a reliable biosensor. Many strategies have been used including direct adsorption, crosslinking with glutaraldehyde, covalent binding, and entrapment in polymerized films or gels [15]. As carrier material in immobilizing enzyme, conducting polymers have been receiving great and broad interests. The immobilization of an enzyme into an electropolymerized film offers many attractive features since the process is instrumentally controlled and an enzyme electrode can be easily prepared in a rapid one-step procedure. The most popular conducting polymers for the immobilization of enzyme are polypyrrole (PPy) [16, 17], polyaniline [18] and polythiophene Especially, PPy and its derivatives are most widely used for entrapping enzyme, because PPy can be easily electrodeposited onto an electrode surface from aqueous solutions, which are compatible with most biological elements [15, 19].

Many researches indicated that the enzyme can be entrapped by PPy matrix immobilized on the surface of Pt, Au and glassy carbon (GC) electrodes. But electropolymerization on the surface of screen-printed carbon paste electrodes has been few reported. Screen-printing technology is a low-cost technology which allows depositing thick films (a few to hundreds of micrometers) and is well suited for mass production and portable devices [20]. They allow fast and easy monitoring. For example, disposable screen-printed enzyme strips are widely used by diabetic patients for self-monitoring of their blood glucose levels [21]. Such a microfabrication route offers high-volume production of extremely inexpensive and yet highly reproducible disposable enzyme electrodes.

In this paper, the fabrication of a disposable hydrogen peroxide amperometric biosensor based on screen-printing technology was described. Carbon and silver ink were used to form electrodes while silver was employed as conductive lead. HRP was immobilized in a PPy film electropolymerized on the surface of screen-printed carbon paste electrodes. Potassium ferrocyanide was deposited as electron transfer mediator to facilitate efficient electron transfer. The biosensor can determine hydrogen peroxide concentration using only 1 µL sample. Different operational parameters influencing the biosensor response, i.e. monomer, electrolyte and HRP concentrations, PPy film growth rate and thickness were evaluated and optimized.

2. EXPERIMENTAL

2.1. Materials and apparatus

HRP (>300 U/mg) was purchased from Qiude Biochemical Engineering CO., Ltd., China. Hydrogen peroxide (30%, w/w), pyrrole (98%), potassium ferrocyanide, lithium perchlorate ($LiClO_4$) and other chemicals were of analytical grade without further purification. Carbon paste (Jelcon CH-10) was purchased from Jujo Chemical CO., Ltd., Japan. Double-distilled water was used in all experiments.

0.25 M hydrogen peroxide standard solution was prepared every three days and stored at room temperature without light. Diluted hydrogen peroxide standard solutions were freshly prepared directly prior to use, carried out in a 0.2 M phosphate buffered saline (PBS) solution (pH 7.0) containing 0.2M NaCl.

Amperometric measurements were performed by a CHI760B Electrochemical Work Station (CH Instrument Inc., USA). During the electropolymerization, a platinum disk (2 mm^2) and Ag/AgCl were used as counter and reference electrodes, respectively. The electrodes were screen-printed using a MT-750A screen printing machine (Ming Tai Screen Printing Machine CO., Ltd., Taiwan)

2.2. Electrode preparation

Screen-printed electrodes were fabricated on polypropylene (PP) sheets using the screen-printer. Three masks were used to form sliver conducting lead wires, carbon paste film for electrochemical reaction and insulating film, the process has been described in detail previously [22]. The structure of the biosensor is shown in Figure 1, consisting of an Ag/AgCl pseudo-reference electrode and two carbon electrodes, which acting as working and counter electrodes, respectively. The function of the silver leads was to improve electric conductivity of the electrodes. The reaction area was defined by the insulating film covering on the carbon paste film. The size of working electrode in reaction area was 1 mm×2 mm while the sensor substrate was 35 mm×10 mm.

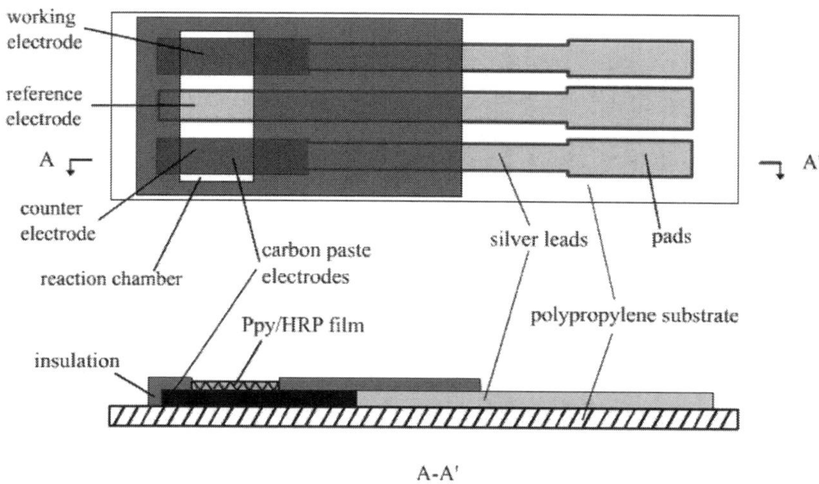

Figure 1. The structure of the screen-printed electrodes, the sensor substrate was 35 mm×10 mm while the working electrode area was 1 mm×2 mm.

2.3. Fabrication of enzyme electrode

The hydrogen peroxide biosensor was fabricated by electropolymerization in a 5 mL 0.2 M PBS solution (pH 7.0) containing 0.075 M pyrrole, 0.075 M $LiClO_4$ and 0.8 mg/mL HRP. One of the screen-printed carbon paste electrodes was used as working electrode while the platinum disk and Ag/AgCl were used as counter and reference electrodes, respectively. Electrochemical polymerization carried out potentiostatically at +1 V vs. Ag/AgCl until a final optimum charge deposit (50 mC/cm^2) was reached. The electrodes were washed and rinsed in double-distilled water to remove the unfixed enzyme and then were dried at room temperature. Subsequently 1 μL PBS solution (pH 7.0) containing 0.1 M potassium ferrocyanide was coated on reaction area uniformly and dried at room temperature in dark. For comparison, a biosensor was fabricated in a similar way except the HRP (2 mg/mL) was immobilized by physical adsorption.

2.4. Measurement

Amperometric measurements were carried out using a CHI760B electrochemical workstation. The working potential was -300 mV vs. the screen-printed Ag/AgCl pseudo-reference electrode, 1 μL hydrogen peroxide standard solution was dropped onto reaction area of enzyme electrode uniformly. Current-time curves of the amperometry were recorded using an IBM PC compatible computer via a RS232 series port communicating to the electrochemical workstation at room temperature. The response time of the sensors, time needed to reach a plateau corresponding to the steady state when the testing sample was added, was 40 seconds. The response current to H_2O_2 was determined by subtracting the background current from the observed current. The calibration curve was obtained with testing samples of different hydrogen peroxide concentration to investigate the characteristics of the biosensor to determine hydrogen peroxide concentration.

3. RESULTS AND DISCUSSIONS

3.1. Optimization of the electropolymerization

The sensitivity of a hydrogen peroxide biosensor depends on the activity of the immobilized HRP enzyme in the PPy film. The activities of the immobilized HRP enzyme were electrochemically analyzed.

In the presence of 1 µL 1 mM H_2O_2 solution, the effects of pyrrole concentration, PPy film thickness, electropolymerization potential, $LiClO_4$ concentration and HRP concentration on the response of the hydrogen peroxide biosensor were evaluated.

3.1.1. Pyrrole concentration

In potentiostatic electropolymerization mode, pyrrole monomer concentration is an important factor which influences the development of the polymer. A constant potential of +1.0 V vs Ag/AgCl was applied to achieve the charge deposit of 50 mC/cm^2, while various concentrations of pyrrole were used. Figure 2 shows the response current to 1 mM H_2O_2 of the sensors with the film electropolymerized using pyrrole of different concentration. Lower pyrrole concentration did not allow sufficient polymer formation and HRP entrapment onto the electrode surface. Optimized pyrrole concentration, 0.075 M, was selected in the experiments.

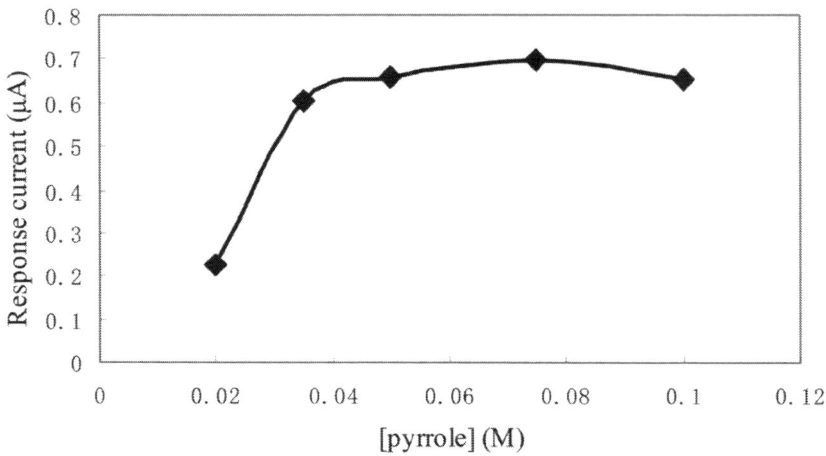

Figure 2. Effect of pyrrole concentration on response current to 1 mM H_2O_2 (1 μL, -300 mV potential applied).

3.1.2. Density of charge deposit

The thickness of the PPy film was determined by the charge deposited during electropolymerization. A thick PPy film could entrap much enzyme, but retard the analyte diffusion. In the research on a PPy glucose biosensor, the PPy film has been proven to act as a serious diffusion barrier, both the diffusion rates of glucose to the immobilized GOx and H_2O_2 to the platinum electrode were retarded severely by the polymer film [23]. However on hydrogen peroxide biosensors, Tatsuma indicated that the PPy film functioned as a conductive material, a part of the electrode material, and didn't retard the transport of the mediators [7]. Our experiment results are shown in Figure 3. When the charge deposited less than 25 mC/cm^2, the response current (curve c) was very small because too little enzyme was entrapped. In the range from 50 to 150 mC/cm^2, the response current to H_2O_2 changed due to the background current (curve b) increasing significantly though the observed current (curve a) remained almost unchanged. It was observed that although the thickness of PPy did not retard the transport of the mediators, it affected the amount of enzyme entrapped within the film and the background current to change the sensitivity of the sensor. In our experiments, the density of charge deposit of 50 mC/cm^2 was selected.

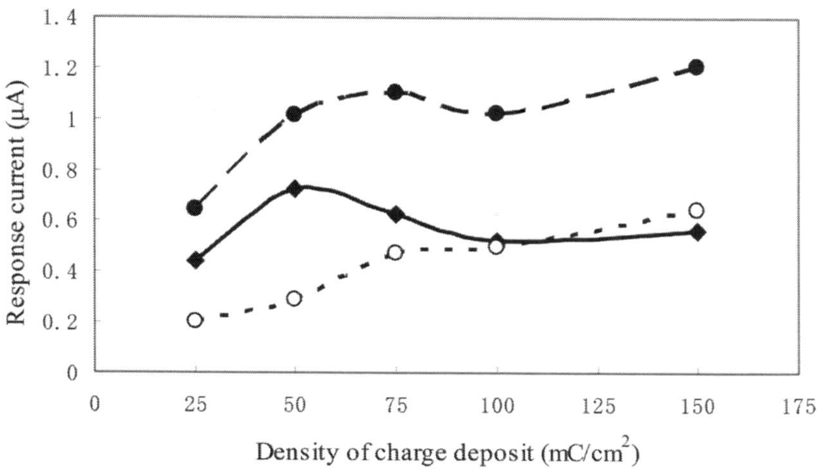

Figure 3. Effect of the density of charge deposit on (a) observed current; (b) background current and (c) response current to 1 H_2O_2 (1 μL, -300 mV potential applied).

3.1.3. Potential for electropolymerization

The effect of electropolymerization potential on enzyme electrode response was evaluated because the potential determined the growth rate of polymer film. With the increase of potential from 0.8 V to 1.2 V, the peak current during electropolymerization process increased from 18 to 175 μA to affect the sensitivity of the sensor, so that the electropolymerization potential should be optimized. When hydrogen peroxide concentration was 1 mM, the potential dependence of response current was shown in Figure 4. The results indicated that the maximum response current was obtained when electropolymerization potential was 1.0 V, which was accordingly used in our experiments.

3.1.4. $LiClO_4$ concentration

The electrolyte concentration should be high enough to facilitate the charge transfer in solution, but it should not compete with the pyrrole monomer cation radicals. Both $LiClO_4$ and KCl were commonly used as electrolyte for pyrrole polymerization [6, 8, 24, 25]. Comparing both electrolytes, we used $LiClO_4$ in the experiments due to its better catalyze effect besides the effect of (inline-equation) doping. When $LiClO_4$ concentration was within a range of 0.075 to 0.15 M, the electropolymerized enzyme sensor performed well, so the $LiClO_4$ concentration of 0.075 M was selected for our experiments.

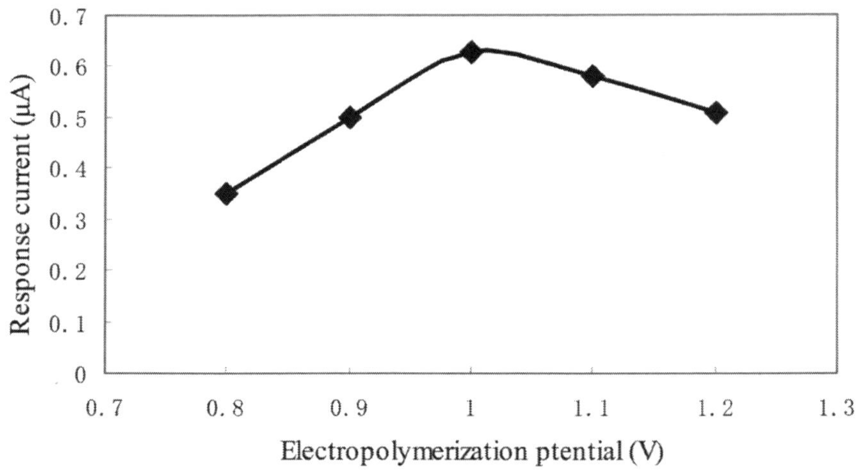

Figure 4. Effect of electropolymerization potential on response current to 1 mM H_2O_2 (1 μL, -300 mV potential applied).

3.1.5. Enzyme concentration

The HRP concentration in the pyrrole monomer solution for preparing the enzyme electrode was a key factor to affect the sensitivity, because the sensor response depended upon the amount of immobilized HRP. Figure 5 shows the response current to 1 mM H_2O_2 of the sensors using HRP of different concentration. With the increase of HRP concentration from 0.2mg/ml to 0.8mg/ml, the response current was almost doubled. When the HRP concentration was higher than 0.8 mg/ml, the response current of the sensor did not increase further. So the HRP concentration of 0.8mg/ml was used in the experiments for the economic consideration.

3.2. pH and electrolyte influences on biosensor response

The pH influence was investigated by amperometric measurement of 1 mM H_2O_2 in PBS 0.067 M at different pH values between 5.0 and 9.0 as shown in Figure 6. The maximum response current was observed at pH 7.0 in agreement with [Ref.26]. In order to obtain the maximum bioactivity and optimal sensitivity, PBS solution of pH 7.0 was selected for our experiments.

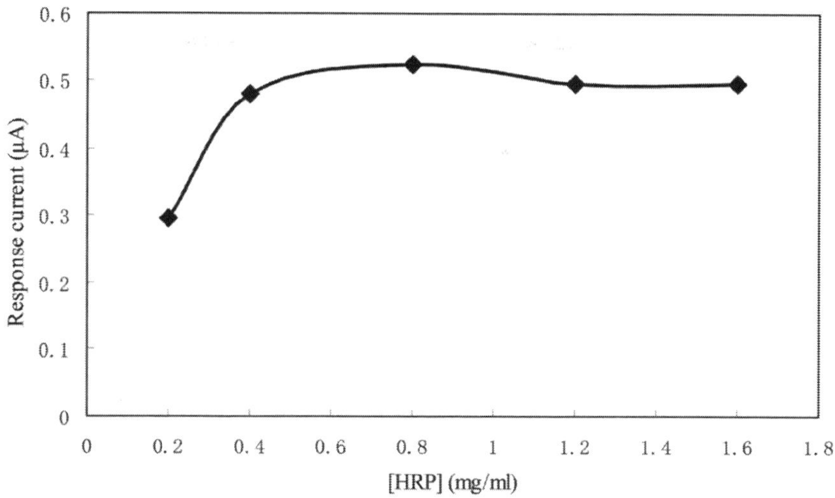

Figure 5. Effect of HRP concentration on response current to 1 mM H_2O_2 (1 μL, -300 mV potential applied).

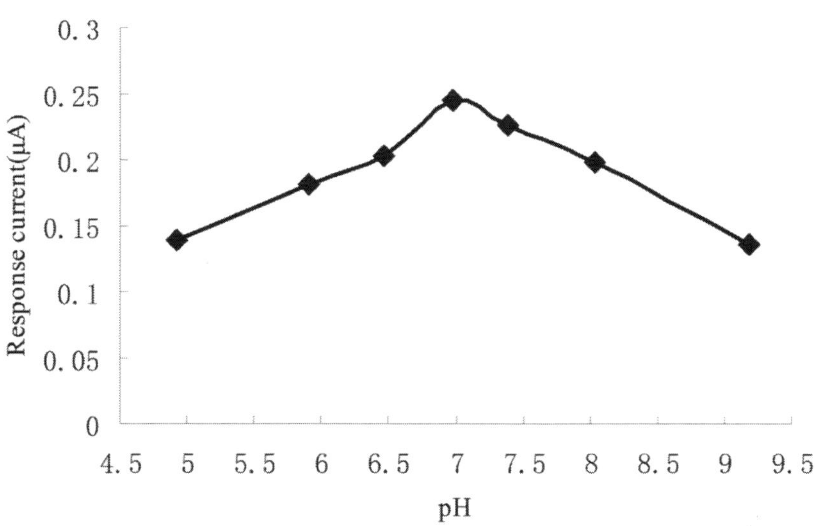

Figure 6. Effect of pH on response current to 1 mM H_2O_2 (1 μL, -300 mV potential applied).

Phosphate (0.033, 0.067, 0.1, 0.2 M) was chosen to study the influence of electrolyte concentration. The results showed that the effect of phosphate concentration on the response was little. So 0.2 M PBS solution was selected in the subsequent experiments.

3.3. Effect of ferrocyanide on the response current of the biosensor

The ferrocyanide improves the charge transfer between the redox active site of the enzyme and the electrode, and should exhibit a fast electron exchange rate, while their electrochemical transformation usually takes place at low potentials to avoid the interference of other electroactive species normally present (e.g. ascorbic acid and uric acid). Figure 7 shows the cyclic voltammetry of the electrode in the absence (B) and presence (A) of ferrocyanide. Comparing Figures 7(A) and 7(B), it is obviously that with the presence of mediator, the reduce reaction took place at lower potentials (0 V vs. -0.5 V). From Figure 7(B), it indicated that without the ferrocyanide as mediator, biosensor response was likely caused by the direct electron transfer between HRP and PPy with subsequent reduction of the oxidized PPy. It's similar to those reported in previous literatures [6, 7].

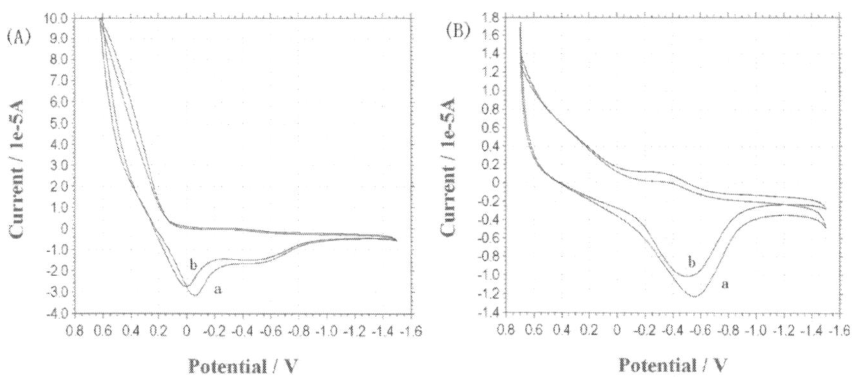

Figure 7. Cyclic voltammetry of the electrodes in the PBS solution (pH7.0). (A) electrodes modified with PPy/HRP/ ferrocyanide (B) electrodes modified with PPy/HRP. (curve a-- with 1 mMH_2O_2, curve b-- absence of H_2O_2.)

3.4. Sensor characteristics

The calibration curve of the sensor is shown as curve (a) in Figure 8. Eight hydrogen peroxide samples of different concentrations were measured using the biosensor fabricated under the optimized condition as described, and each sample of a certain concentration was measured five times. The inset illustration indicates the linear range of the PPy-HRP electrode response was from 0.1 to 2 mM, and the sensitivity corresponding to the linear range was about 33.24 μA mM^{-1} cm^{-2}. The equation of the response was [Y(μA)=0.6647X(mM)-0.0234, R^2=0.9953].

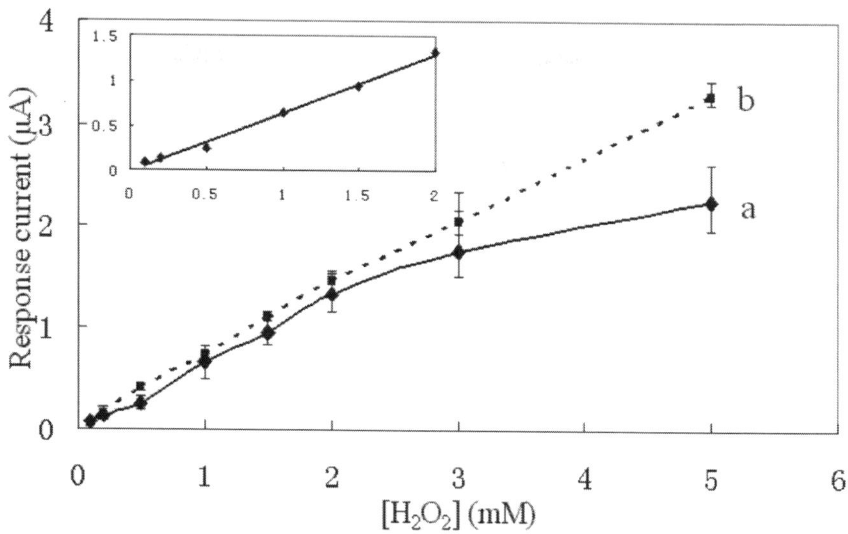

Figure 8. Calibration curve of H_2O_2 biosensor with HRP immobilized by (a) PPy film and (b) physical adsorption. The inset indicates calibration curve obtained for the linear range. Experimental condition: 1 μL H_2O_2, -300 mV applied potential.

The reproducibility between electrodes represented by the relative standard deviation (R.S.D.) was measured. For all the H_2O_2 concentrations investigated, the average R.S.D. was 10.24% (n=5).

The calibration curve of a sensor with HRP immobilized by physical adsorption is shown as curve (b) in Figure 8. Although the response of the electropolymerized sensor was a little lower than the one of the sensor with physically adsorbed HPR, the quantity of the HRP fixed in the PPy film can be relative easily adjusted by quantitatively controlling the electrical charge of polymerization.

Different from previous researches reported, we electropolymerized PPy-HRP on a screen-printed carbon paste electrode rather than Pt or GC electrode [6, 27]. The sensor was developed for disposable use to allow quick and easy hydrogen peroxide monitor while only 1 μL sample was required. The sensor exhibited a sensitivity of 33.24 μA mM^{-1} cm^{-2} which was higher than the hydrogen peroxide sensors reported in [Ref.6, 27] and allowed to be measured with a handheld meter conveniently.

3.5. Sensor lifetime

The operational stability was examined by measuring the response to 1 mM H_2O_2. The sensors were stored in 0.2 M PBS (pH 7.0) at 5 Celsius degree for 3 weeks, while measurements were conducted every 2 days during the first week and then once a week subsequently. The response did not change significantly.

4. CONCLUSIONS

A hydrogen peroxide biosensor with PPy-HRP electropolymerized on a screen-printed carbon paste electrode was studied. When the condition for electropolymerization was optimized, the sensor exhibited a sensitivity of 33.24 $\mu A\ mM^{-1}\ cm^{-2}$ to H_2O_2 with a linear range of 0.1 to 2 mM while only 1 µL sample was required. This method is promising for manufacture of economic disposable hydrogen peroxide biosensors.

ACKNOWLEDGMENTS

This work has been funded by National Natural Science Foundation of China Grant #60421002.

REFERENCES

1. Ruzgas, T.; Csoregi, E.; Emneus, J.; Gorton, L.; Marko-Varga, G. Peroxidase-modified electrodes: fundamentals and application. Anal. Chim. Acta **1996**, 330, 123–138.
2. Hurdis, E.C.; Hendrik Romeyn, J. Accuracy of determination of hydrogen peroxide by cerate oxidimetry. Anal. Chem. **1954**, 26, 320–325.
3. Matsubara, C.; Kawamoto, N.; Takamura, K. Oxo[5, 10, 15, 20-tetra(4-pyridyl)porphyrinato] titanium(IV): an ultra-high sensitivity spectrophotometric reagent for hydrogen peroxide. Analyst **1992**, 117, 1781–1784.
4. Aizawa, M.; Ikariyama, Y.; Kun, H. Photovoltaic determination of hydrogen peroxide with a biophotodiode. Anal. Lett. **1984**, 17, 555–564.
5. Nakashima, K.; Maki, K.; Kawaguchi, S.; Akiyama, S.; Tsukamoto, Y.; Imai, K. Peroxyoxalate chemiluminescence assay of hydrogen peroxide and glucose using 2,4,6,8-tetrathio-morpholinopyrimido[5,4-d]pyrimidine as a fluorescent component. Anal. Sci. **1991**, 7, 709.

6. Serradilla Razola, S.; Lopez Ruiz, B.; Mora Diez, N.; Mark, H.B., Jr; Kauffmann, J.M. Hydrogen peroxide sensitive amperometric biosensor based on horseradish peroxidase entrapped in a polypyrrole electrode. Biosens. Bioelectron. **2002**, 17, 921–928.

7. Tatsuma, T.; Gondaira, M.; Watanabe, T. Peroxidase-incorporated polypyrrole membrane electrodes. Anal. Chem. **1992**, 64, 1183–1187. [Google Scholar]

8. Thanachasai, S.; Rokutanzono, S.; Yoshida, S.; Watanabe, T. Novel hydrogen peroxide sensors based on peroxidase-carrying poly{pyrrole-co-[4-(3-pyrrolyl)butanesulfonate] }copolymer films. Anal. Sci. **2002**, 18, 773–777.

9. Oungpipat, W.; Alexander, P.W.; Southwell-Keely, P. A reagentless amperometric biosensor for hydrogen peroxide determination based on asparagus tissue and ferrocene mediation. Anal. Chim. Acta **1995**, 309, 35–45.

10. Wang, L.; Wang, E. A novel hydrogen peroxide sensor based on horseradish peroxidase immobilized on colloidal au modified ITO electrode. Electrochem. Commun. **2004**, 6, 225–229.

11. Yang, Y.; Mu, S. Determination of hydrogen peroxide using amperometric sensor of polyaniline doped with ferrocenesulfonic acid. Biosens. Bioelectron. **2005**, 21, 74–78.

12. Vidal, J.-C.; Garcia-Ruiz, E.; Castillo, J.-R. Recent Advances in Electropolymerized Conducting Polymers in Amperometric Biosensors. Microchim. Acta **2003**, 143, 93.

13. Vidal, J.C.; Garcia, E.; Castillo, J.R. Development of a platinized and ferrocene-mediated cholesterol amperometric biosensor based on electropolymerization of polypyrrole in a flow system. Anal. Sci **2002**, 18, 537–542.

14. Hong, M.Y.; Chang, J.Y.; Yoon, H.C.; Kim, H.S. Development of a screen-printed amperometric biosensor for the determination of L-lactate dehydrogenase level. Biosens. Bioelectron. **2002**, 17, 13–18. [Google Scholar]

15. Cosnier, S. Biosensors based on electropolymerized films: new trends. Anal. Bioanal. Chem. **2003**, 377, 507–520.

16. Xue, H.; Mu, S. Bioelectrochemical response of the polypyrrole xanthine oxidase electrode. J. Electroanal. Chem. **1995**, 397, 241. [Google Scholar]

17. Koopal, C.G.J.; Bos, A.A.C.M.; Nolte, R.J.M. Third-generation glucose biosensor incorporated in a conducting printing ink. Sens. Actuators B **1994**, 18, 166.

18. Xue, H.; Shen, Z.; Li, C. Improved selectivity and stability of glucose biosensor based on in situ electropolymerized polyaniline-polyacrylonitrile composite film. Biosens. Bioelectron. **2005**, 20, 2330–2334.

19. Vidal, J.C.; García, E.; Méndez, S.; Yarnoz, P.; Castillo, J.R. Three approaches to the development of selective bilayer amperometric biosensors for glucose by in situ electropolymerization. Analyst **1999**, 124, 319–324.

20. Hart, J.P.; Crew, A.; Crouch, E.; Honeychurch, K.C.; Pemberton, R.M. Some recent designs and developments of screen-printed carbon electrochemical sensors/biosensors for biomedical, environmental, and Industrial analyses. Anal. Lett. **2004**, 37, 789–830.

21. Newman, J.D.; Turner, A.P.F. Home blood glucose biosensors: a commercial perspective. Biosens. Bioelectron. **2005**, 20, 2435–2453.

22. Li, G.; Ma, N.Z.; Wang, Y. A new handheld biosensor for monitoring blood ketones. Sens. Actuators B **2005**, 109, 285–290.

23. Uang, Y.M.; Chou, T.C. Criteria for designing a polypyrrole glucose biosensor by galvanostatic electropolymerization. Electroanalysis **2002**, 14, 1564–1570.

24. Da Silva, S.; Shan, D.; Cosnier, S. Improvement of biosensor performances for nitrate determination using a new hydrophilic poly(pyrrole-viologen) film. Sens. Actuators B **2004**, 103, 397–402.

25. Vela, M.H.; De Jesus, D.S.; M., C.M.C.; Araujo, A.N.; Montenegro, M.C.B.S.M. Electroimmobilization of MAO into a polypyrrole film and its utilization for amperometric flow detection of antidepressant drugs. Electroanalysis **2003**, 15, 133–138.

26. Harbury, H.A. Oxidation-reduction potentials of horseradish peroxidase. J. Biol. Chem. **1957**, 225, 1009.

27. Gaspar, S.; Habermuller, K.; Csoregi, E.; Schuhmann, W. Hydrogen peroxide sensitive biosensor based on plant peroxidases entrapped in Os-modified polypyrrole films. Sens. Actuators B **2001**, 72, 63–68.

Chapter 10

Voltammetric Behaviour of Sulfamethoxazole on Electropolymerized-Molecularly Imprinted Overoxidized Polypyrrole

Sabriye Perçin Özkorucuklu [1], Yücel Şahin [2,*] and Güleren Alsancak [1]

[1] *Süleyman Demirel University, Faculty of Arts and Sciences, Department of Chemistry, 32260 Isparta, Turkey*

[2] *Anadolu University, Faculty of Science, Department of Chemistry 26470, Eskişehir, Turkey*

ABSTRACT

In this work, preparation of a molecularly imprinted polymer (MIP) film and its recognition properties for sulfamethoxazole were investigated. The overoxidized polypyrrole (OPPy) film was prepared by the cyclic voltammetric deposition of pyrrole (Py) in the presence of supporting electrolyte (tetrabutylammonium perchlorate-TBAP) with and without a template molecule (sulfamethoxazole) on a pencil graphite electrode (PGE). The voltammetric behaviour of sulfamethoxazole on imprinted and non-imprinted (NIP) films was investigated by differential pulse voltammetry (DPV) in Britton-Robinson (BR) buffer solutions prepared in different ratio of acetonitrile-water binary mixture, between the pH 1.5 and 7.0. The effect of the acetonitrile-water ratio and pH, monomer and template concentrations, electropolymerization cycles on the performance of the MIP electrode was investigated and optimized. The MIP electrode exhibited the best reproducibility and highest sensitivity. The results showed that changing acetonitrile-water ratio and pH of BR buffer solution changes the oxidation peak current values. The highest anodic signal of sulfamethoxazole was obtained in BR buffer solution prepared in 50% (v/v) acetonitrile-water at pH 2.5. The calibration curve for sulfamethoxazole at MIP electrode has linear region for a concentration range of 25.10^{-3} to 0.75 mM (R^2=0.9993). The detection limit of sulfamethoxazole was found as $3.59.10^{-4}$ mM (S/N=3). The same method was also applied to determination of sulfamethoxazole in commercial pharmaceutical samples. Method precision (RSD<1%) and recoveries (>87%) were satisfactory. The proposed method is simple and quick. The polypyrrole (PPy) electrodes have low response time, good mechanical stability and are disposable simple to construct.

Keywords Molecularly imprinted polymer; overoxidized polypyrrole; pencil graphite electrode; electropolymerization; sulfamethoxazole; sulfonamides

1. INTRODUCTION

Sulfonamides have been used as antibacterial agents for last 60 years. A number of them have found widespread use in animal husbandry and, to a lesser extent, in the treatment of human infections such as bronchitis and urinary tract infections. The use of sulfonamide drugs in veterinary care without a proper withdrawal period will cause accumulation of sulfonamides in meat, eggs and milk as well as in fish. Because of the possible risk of resistance development in humans, the European Union (EU) has set a maximum residue limit (MRL, 0.1 µg/g) for sulfonamides of animal originated products [1-2].

These low concentration limits led to development of fast and sensitive method to screen foodstuffs for sulfonamide based drugs. Several techniques have been reported in the literature for the determination of sulfonamides, i.e. gas chromatography (GC) and gas chromatography-mass spectrometry (GC-MS) [3], high performance liquid chromatography [4-7] and capillary electrophoresis [8-9] using various detectors. However there are only a few reports available on the electrochemical properties of sulfonamides [10-12].

Over the past two decades, molecular imprinted polymers (MIPs) have attracted a broad interest from scientists engaged in sensor development. This can be explained by the serious potential advantages of using MIPs in place of natural receptors and enzymes for their superior stability, low cost and easy preparation. The general principal of molecular imprinting is based on a process where functional and cross-linking monomers are copolymerization in presence of a target analyte (the imprint molecule) which acts as a molecular template. This procedure can be accomplished via either reversible covalent bonding or non-covalent interactions between monomers and imprint molecules. Other preparation methods of molecular imprinting polymers are chemical grafting, soft lithography technique [13], molecular self-assembled approach [14] and electropolymerization [13]. These films can also be synthesized in situ at an electrode surface by electropolymerization technique.

This technique has some attractive features including the easy adherence of the polymeric films to the surface of conducting electrodes of any shape and size and the ability to control thickness of the films under different depositions conditions [15]. Various types of electrosynthesized polymers based on molecular imprinting have been reported in the literature including poly(o-phenylenediamine) [15], poly(2-mercaptobenzimidazole) [16], polypyrrole [17], polyphenol [18] and copolymer of aniline with o-phenylendiamine [19].

Among various types of conducting polymers, polypyrrole has many attractive features as a molecular recognition system, since it can be used in a neutral pH region, and its stable films can conveniently be polymerized on various substrate materials. Sensors based on molecularly imprinted polypyrrole for amino acids [20], caffeine [21-23], paracetamol [24], ascorbic acid [25], isoniazid [26], glycoproteins [27] were reported. Our group prepared a molecularly imprinted polypyrrole-based films for the determination of paracetamol and ascorbic acid. The performance of the imprinted films was evaluated by differential pulse voltammetry. This film exhibited a high selectivity and sensitivity toward paracetamol and ascorbic acid [24-25]. PPy undergoes overoxidation at positive potentials, and this process has often been regarded as an undesirable degradation process, which leads to the loss of conductivity and dedoping [28]. Despite these disadvantages, OPPy has been used in some electroanalytical applications; the overoxidized film works as a porous electrode coating, which has cation-exchange and molecular sieve properties [20, 26, 29-33]. It has been reported that, during oxidation polypyrrole loses its electroactivity due to ejection of dopant, and oxygen-containing groups such as carbonyl and carboxyl are introduced to the pyrrole unit [34]. The accumulative properties of the film for cationic species might be attributed to the introduction of carboxylate. The best results are achieved if during electrochemical deposition OPPy is imprinted by small molecular weight molecules [21-22]. Moreover, attempts to imprint PPy by large molecular weight rigid structure possessing proteins were reported as well as, in this case viral envelope proteins possessing rigid structure were imprinted within OPPy [27]. PPy is overoxidized chemically or electrochemically to cause dedoping as shown in Scheme 1.

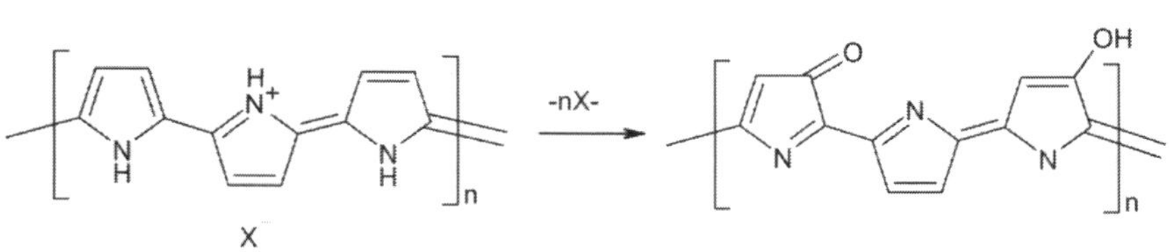

Scheme 1. The overoxidation reaction of polypyrrole

The use of a differential pulse voltammetry to determine the sulfamethoxazole using pencil graphite electrode coated with overoxidized polypyrrole films by imprinting electropolymerization was reported for the first time in this work. The structure of sulfamethoxazole is shown in Figure 1. Its successful application for determination of this drug in pharmaceutical sample has been demonstrated.

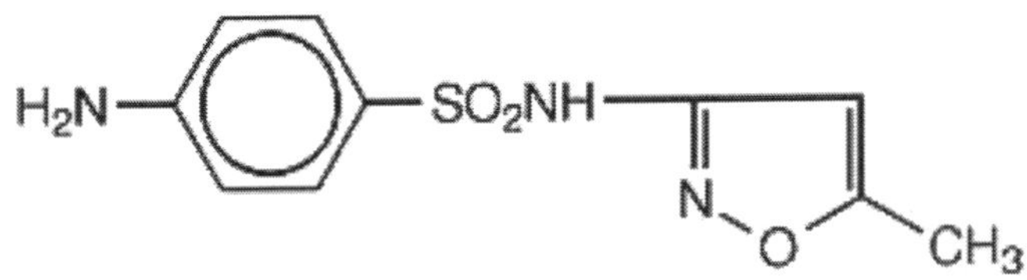

Figure 1. The structure of sulfamethoxazole.

2. RESULTS AND DISCUSSION

2.1. Electropolymerization of molecularly imprinted overoxidized polypyrrole

The electrochemical behaviour of pyrrole was investigated in acetonitrile solution of 0.1 M TBAP using potential cycling between -0.6 and +1.40 V (versus Ag/AgCl) with PGE. Electrooxidation of pyrrole monomer occurs at the anode and the resulting polymer deposits onto the surface of PGE. An anodic peak of pyrrole was observed at a peak potential of 1.20 V. The corresponding reduction process was not observed on the cyclic voltammogram. The oxidation peak corresponds to the formation of pyrrole radical cations. Figure 2a demonstrates five cycles obtained in the same solution. The formation and growth of the polymer film can be easily seen in this figure. The peaks due to the oxidation and reduction of the film increase intensity as the film grows. A broad oxidation peak was observed at the peak potential of +0.15 V and reverse cathodic peak was seen at a peak potential of -0.10 V.

For imprinted electropolymerizations, 5 mM sulfamethoxazole as a template was added to the electrochemical cell. Figure 2b demonstrates the cyclic scans of electropolymerization of pyrrole in the presence of sulfamethoxazole. The effect of sulfamethoxazole on the electropolymerization of pyrrole can be seen easily in this figure. Even though sulfamethoxazole is an electro-inactive template, the oxidation peak potential of polypyrrole shifted to more anodic potentials, from 0.15 to 0.35 V, and approximately ten times smaller oxidation peak current value was observed. This oxidation peak indicates that the template is becoming part of the polymeric chain [17]. Sulfamethoxazole diffuses towards the surface of PGE entrapped in the polymer matrix during polymerization process of pyrrole. This is the key process towards achieving a voltammetric sensor based on molecular imprinting polymers. The creation of the molecular imprint is favored by the diffusion of the electroactive template, generating a far higher number of recognition sites than those previously obtained with non-electroactive template. If the template is non-electroactive, molecular diffusion towards the electrode is interrupted after the first scans by the formation of the non-conductive polymeric layer, which prevents the template forming part of this layer and thereby decreasing the peak current intensity.

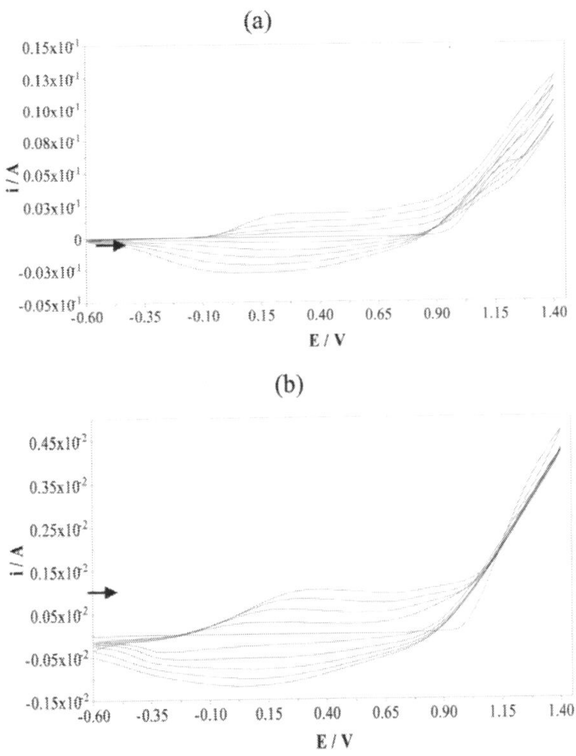

Figure 2. Cyclic voltammograms taken during the electropolymerization of pyrrole (0.1 M). (a) Multisweep cyclic voltammograms without and (b) with sulfamethoxazole (5 mM) onto a pencil graphite electrode (scan rate: 100 mVs^{-1}; supporting electrolyte: 0.1 M TBAP; number of scans: 5)

During the electrodeposition of pyrrole, sulfamethoxazole template molecules are trapped in to the polymer matrix as a result of the ability of these molecules to interact with the pyrrole units. Figure 3. shows a schematic representation of imprinting and removal of sulfamethoxazole from sulfamethoxazole imprinted polypyrrole modified pencil graphite electrode. The oxygen atom in the S=O group of the sulfamethoxazole molecule forms a hydrogen bond with the hydrogen atom in the N-H group of the pyrrole units (Figure 3). Hydrogen bonding could occur between the hydrogen in the amino group of sulfamethoxazole structure and the nitrogen atom of the pyrrole N-H groups. Chain branching and cross linking in polypyrrole generate a three-dimensional matrix with niches containing the template sulfamethoxazole. This imprinting process creates a microenvironment for the recognition of sulfamethoxazole molecule based on shape selection and positioning of the functional groups.

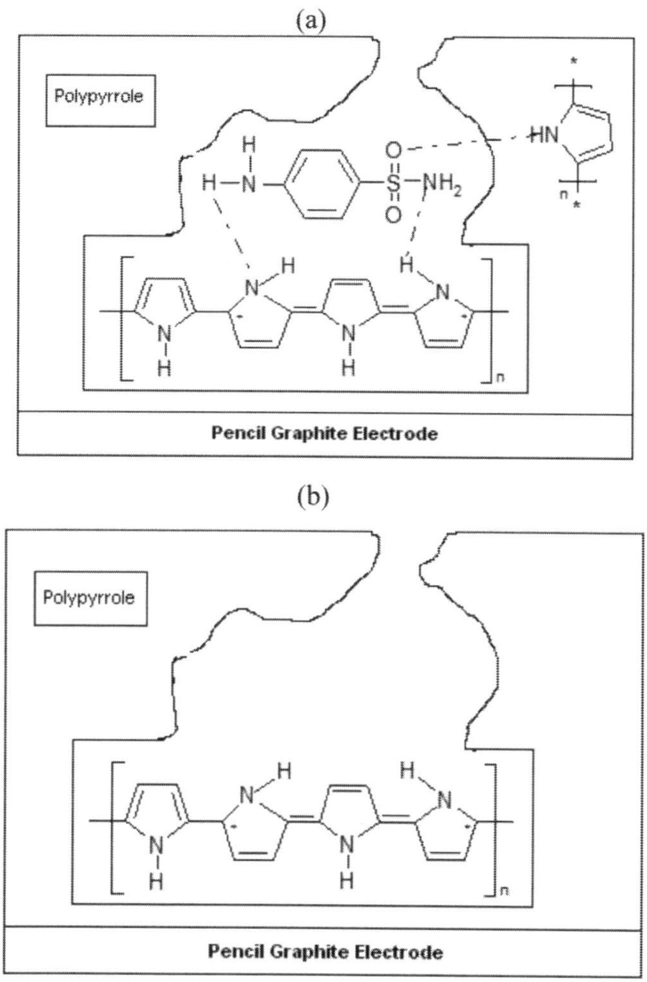

Figure 3. Schematic representation of **(a)** imprinting and **(b)** removal of sulfamethoxazole from sulfamethoxazole imprinted polypyrrole modified pencil graphite electrode.

In order to remove the entrapped template, various strategies are possible: microwave assisted extraction, supercritical fluid desorption, oxidation-reduction of template in polymer, the use of solvent that strongly interacts with polymer causing the swelling of the coating necessary for the template release or overoxidation. To remove the template inside the polypyrrole, overoxidation process was used in our work. The polymer was overoxidized by scanning the potential in a range of +0.80 to +1.20 V during twenty cycles (scan rate: 50 mV/s) in 0.1 M NaOH solution.

2.2. Effect of the acetonitrile-water ratio and pH

The mechanism for electrochemical oxidation of sulfamethoxazole depends on the acetonitrile-water ratio and pH of the solution. The effects of acetonitrile-water ratio and pH of the solution on the MIP electrode is illustrated in Figure 4 by DPV. The highest peak current intensity was observed in %50 (v/v) acetonitrile-water Britton-Robinson buffer at pH 2.5, giving an oxidation peak at 1.15 V.

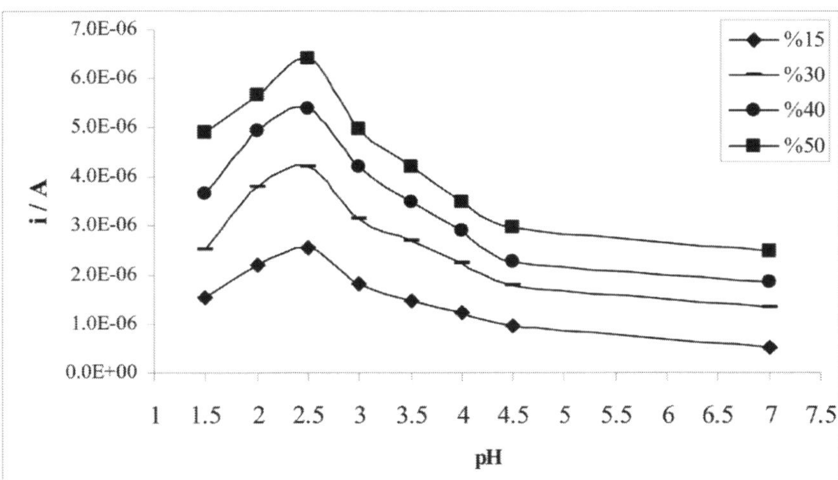

Figure 4. Effects of acetonitrile-water ratio and pH on the MIP modified PGE response for 0.10 mM sulfamethoxazole in Britton-Robinson buffer solution.

2.3. Electrochemical behaviour of sulfamethoxazole

The electrochemical behaviour of sulfamethoxazole (0.10 mM) was investigated by MIP and NIP electrodes in 50% (v/v) acetonitrile-water BR buffer at pH 2.5 by DPV. The catalytic effect of the MIP electrode is demonstrated in Figure 5. Sulfamethoxazole gives an oxidation peak response at about 1.08 V and 1.15 V (versus Ag/AgCl) at the NIP and MIP electrodes, respectively. The enhanced peak current response and a shift in the oxidation potential of sulfamethoxazole by about 70 mV are a clear evidence of the catalytic effect of the MIP electrode towards the oxidation of sulfamethoxazole.

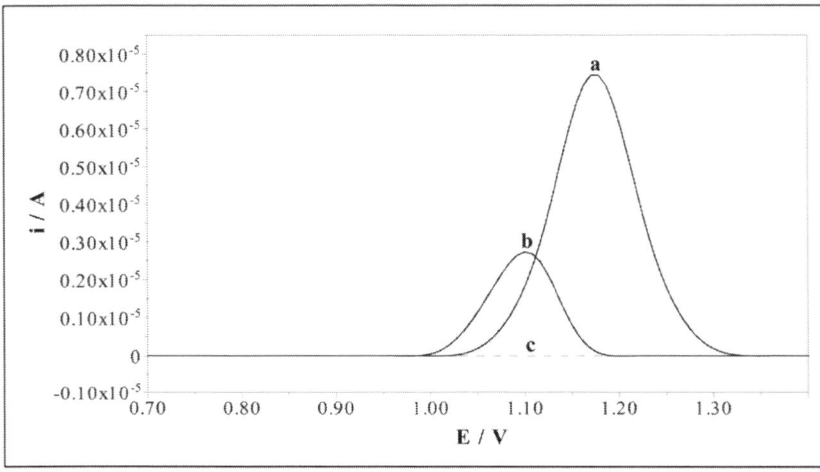

Figure 5. Differential pulse voltammograms for (a) 0.10 mM sulfamethoxazole at MIP electrode (b) 0.10 mM sulfamethoxazole at NIP electrode and (c) blank solution at MIP electrode in 50% (v/v) acetonitrile-water Britton-Robinson buffer at pH 2.5.

The voltammetric determination of sulfamethoxazole using MIP and NIP electrodes results in well-defined concentration dependences. Figure 6 displays differential pulse voltammograms for solutions containing increasing quantities of sulfamethoxazole at the MIP electrode. The calibration curve of sulfamethoxazole at MIP electrode at pH 2.5 was given in Figure 7. Two linear regions were observed between peak current and concentration of sulfamethoxazole for MIP electrode. The first region demonstrates linearity over a concentration range of 25.10^{-3} mM to 0.75 mM with a correlation coefficient of 0.9993. The slope of the second linear region in the concentration range of 0.75-2.0 mM is smaller than the first region's slope. The correlation coefficient of this region was 0.992. Table 1 represents calibration characteristics and related parameters for sulfamethoxazole using MIP electrode. The limit of detections (LOD) and limit of quantification (LOQ) of sulfamethoxazole with MIP electrode were calculated according to the 3s/m and 10s/m criterious, respectively. The results are also shown in Table 1.

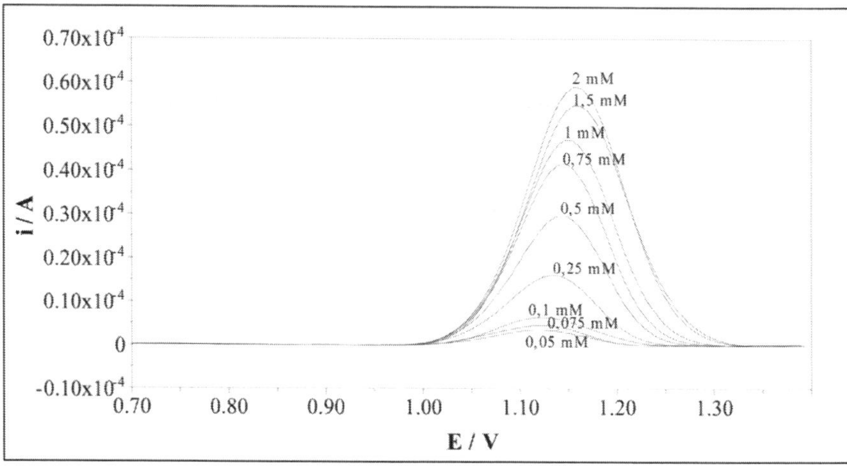

Figure 6. Differential pulse voltammograms of varying concentration at the MIP electrode at pH 2.5.

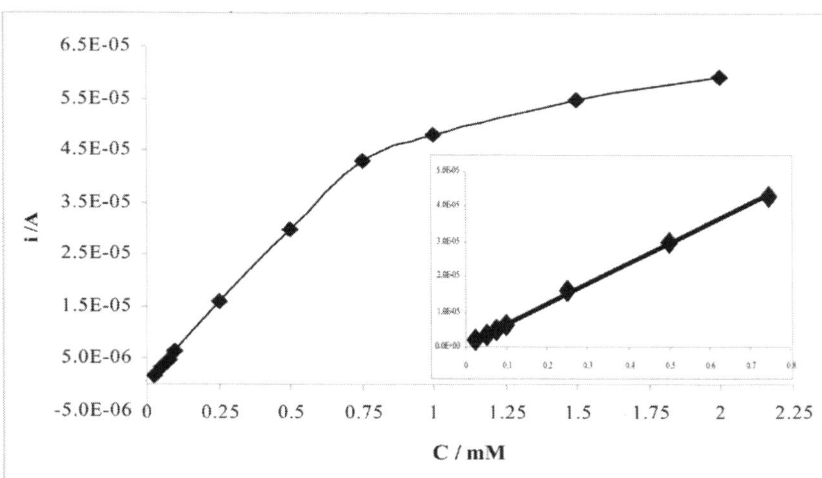

Figure 7. The calibration curves of sulfamethoxazole between peak current and concentration of sulfamethoxazole at MIP electrode.

Table 1. Characteristics of sulfamethoxazole calibration plots.

	Sulfamethoxazole	
Linearity range (mM)	25.10^{-3}- 0.75	0.75-2.0
Slope	$5.75.10^{-5}$	$1.28.10^{-5}$
Intercept	$7.15.10^{-7}$	$3.45.10^{-5}$
Correlation coefficient	0.9993	0.9920
RSD of slope	1.08	0.98
RSD of intercept	0.77	1.81
LOD	$3.59.10^{-4}$ mM	
LOQ	$1.20.10^{-3}$ mM	

2.4. Effect of the monomer concentration

The monomer concentration during polymerization also determined the analytical behaviour of the sensor. To determine the effect of monomer concentration on the response of both MIP and NIP to sulfamethoxazole, the films were grown in solutions of constant concentration of sulfamethoxazole and varying pyrrole concentrations in the range of 25-500 mM by cycling potential between -0.60 V and +1.40 V. Figure 8 shows the variation of the monomer concentration as a function of the current values for sulfamethoxazole. The monomer concentration should be proportional to the thickness of the deposit and amount of imprinted molecule (template) in the polymer matrix. The response of the MIP electrode to sulfamethoxazole was found to increase with increasing pyrrole concentration up to 100 mM. There was considerable decrease in the response of MIP electrode below and above this pyrrole concentration. The current difference between the NIP and NIP electrodes for sulfamethoxazole should be as high as possible. In addition, the signal of the NIP electrode should be a minimum. It can be concluded that the optimum monomer concentration under these conditions was about 100 mM as clearly seen in Figure 8.

2.5. Effect of the electropolymerization cycles

The optimum number of CV cycles to use to form the sensing layer of the electrode was determined from a series of experiments in which electrodes were fabricated with different numbers of cycles (Figure 9).

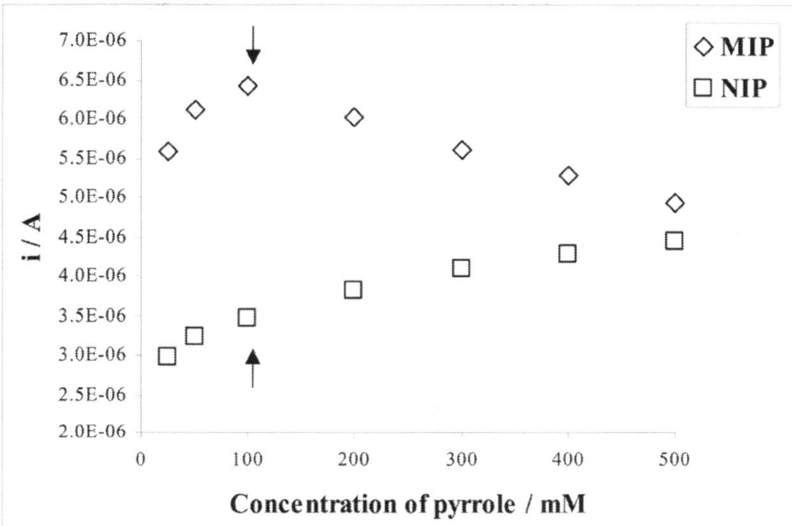

Figure 8. Effect of the monomer concentration on (Δ) MIP and (□) NIP electrodes for 0.10 mM sulfamethoxazole in 50% (v/v; acetonitrile-water Britton-Robinson buffer at pH 2.5.

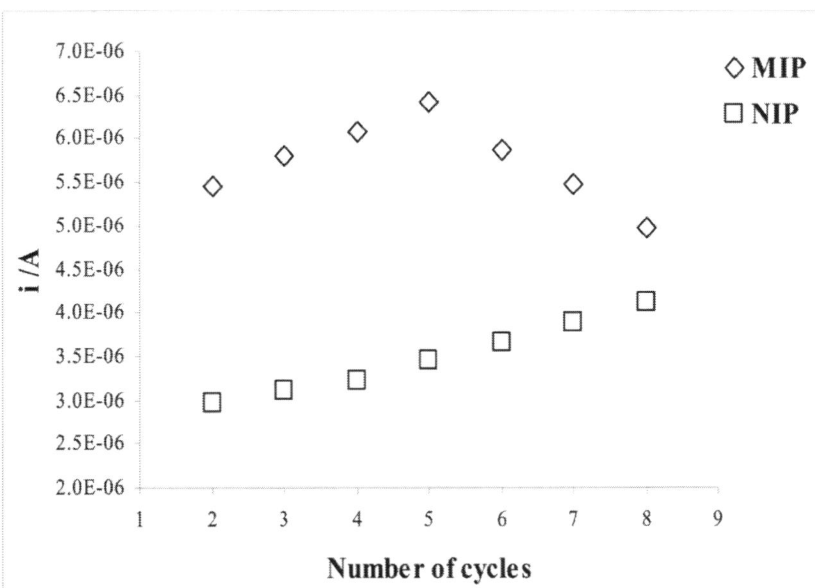

Figure 9. Effect of the number cycle to (Δ) MIP and (□) NIP electrodes for 0.10 mM sulfamethoxazole in 50% (v/v; acetonitrile-water Britton-Robinson buffer at pH 2.5.

The number of cycles applied to the cell during the electropolymerization was found to affect the sensitivity and linearity of the sensor. The MIP electrode produced at lower number of cycles exhibited favorable analytical performance. Higher cycles lead to more extensive electropolymerization, and therefore to the formation of thicker sensing film with less accessible imprinted sites. The response of the NIP electrode to sulfamethoxazole was found to increase with increasing the number of cycles. There was considerable decrease in the performance of the MIP electrode below and above five cycles. The highest current difference between the MIP and NIP electrodes for sulfamethoxazole was obtained by applying five cycles in the electropolymerization. Therefore the optimum polymerization cycles was found to be five.

2.6. Effect of the template concentration

The template concentration at the moment of electropolymerization may also be significant factor in the response of the sensor. Figure 10 shows the effect of the template concentration during electrodeposition of the film in the range of 1.0-10 mM. The response of the MIP electrode to sulfamethoxazole increases with the increase of the template concentration between 1.0 and 5.0 mM. There was a small decrease in the response of MIP electrode above this sulfamethoxazole concentration (5.0 mM). Based on the results, the optimum template concentration was chosen as 5.0 mM.

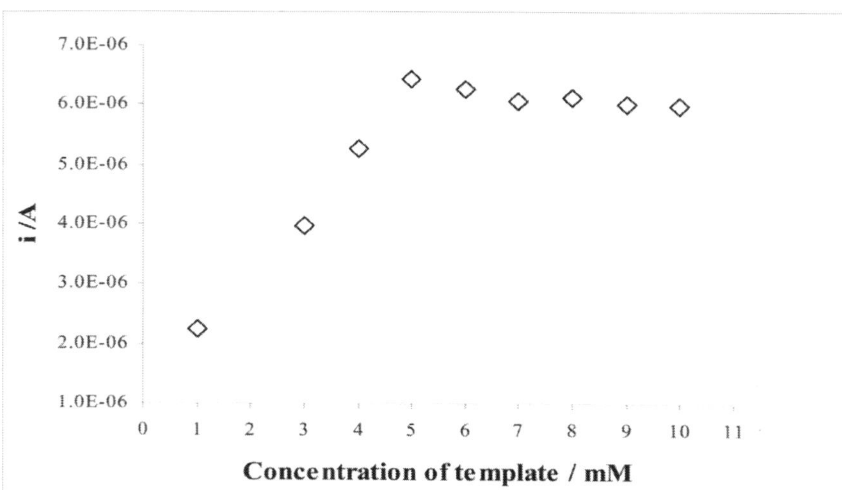

Figure 10. Effect of the template concentration to MIP response.

2.7. Analysis of pharmaceutical samples

In order to demonstrate the practical usage of the biosensor, tablet and syrup having sulfamethoxazole was examined for estimation of sulfamethoxazole by MIP electrode. Solution obtained by dissolution of sulfamethoxazole tablet and syrup were subsequently diluted so that sulfamethoxazole concentration lies in the range of calibration plot. Differential pulse voltammograms were then recorded under exactly identical conditions that were employed while recording differential pulse voltammograms for plotting calibration plot. The amounts of sulfamethoxazole that reported and experimentally determined in tablet and syrup are listed in Table 2. It was found that sulfamethoxazole concentration determined using this method is in good agreement with the manufacturer's stated contents of sulfamethoxazole. To validate the voltammetric detection method, the pharmaceutical samples were also determined with a HPLC method at the optimal chromatographic conditions. Table 2 shows the obtained data from the HPLC and DPV with the MIP electrode. The all values in this table are mean of five replicate measurements. The results reveal that both methods had adequate precision and accuracy. Therefore they both can be applied to the determination of sulfamethoxazole in pharmaceutical formulations.

Table 2. Determination of sulfamethoxazole in pharmaceutical preparations using DPV at molecularly imprinted polypyrrole modified pencil graphite electrode and HPLC.

	DPV		HPLC	
	Trimoks Fort Tablet	Bactrim Syrup	Trimoks Fort Tablet	Bactrim Syrup
Reported Content	0.250[a]	50[b]	0.250[a]	50[b]
Detected Content	0.240[a]	43.7[b]	0.245[a]	47.4[b]
Recovery%	96.0	87.4	98.0	94.8
RSD%	0.90	0.85	1.08	0.76
Student t-test	0.103	0.379	0.420	0.160
F-test	1.500	1.063	$t_{theoretical}$: 2.31	$F_{theoretical}$: 6.39

[a] g/tablet
[b] mg/100 mL

A Student t-test and F-test were carried out on the data to statistically examine the validity of the obtained results. At 95% confidence level, the calculated t and F values were less than of theoretical t and F values showing that there is no significative differences between the DPV and HPLC methods. The developed DPV method is simpler, faster and requires less expensive equipment than chromatographic methods.

2.8. Reproducibility of the MIP electrode

The reproducibility of the molecularly imprinted modified pencil graphite electrode was investigated for 0.10 mM sulfamethoxazole. The peak current response of sulfamethoxazole was determined with five electrodes which produced under the same conditions. The response peak intensity showed a relative standard deviation of 3.4% confirming that the results are reproducible.

3. EXPERIMENTAL SECTION

3.1. Chemicals and reagents

Sulfamethoxazole (>99.9%) was obtained from Sigma-Aldrich (Germany). Acetonitrile (>99.9%, Sigma-Aldrich, Germany), tetrabutylammonium perchlorate (99%, Fluka, Germany), sodium hydroxide (>98%, Merck, Germany), potassium hydrogen phthalate (>99.5%, Merck, Germany, dried at 110 oC before use), phosphoric acid (85%, Riedel-de Haen, Germany), acetic acid (Glacial, Riedel-de Haen, Germany), boric acid (>99.5%, Carlo Erba, Italy) are commercially available as analytical grade reagents. Pyrrole (98%, Aldrich, Germany) was distilled repeatedly under vacuum until a colorless liquid was obtained and kept under nitrogen in darkness at 4°C. Ultra-pure deionized water (Sartorius), with a resistivity of 18.2 MΩ cm, was used for all experiments. Trimoks Fort tablet (Atabay) and Bactrim syrup (Roche) were purchased from a local pharmacy. Freshly prepared solution of sulfamethoxazole was prepared each day owing to its low stability.

3.2. Apparatus

All electrochemical studies were carried out with Autolab PGSTAT 100 Potentiostat/Galvanostat controlled by GPES 4.9 version software (Ecochemie, The Netherlands). The three-electrode system was used for all measurements; a pencil graphite electrode (PGE) as the working electrode, a Pt auxiliary electrode and an Ag/AgCl reference electrode. MA 235 model (Mettler Toledo, Switzerland) pH-Ion meter with InLab 416 Ag/AgCl glass electrode was used for pH measurements. Chromatographic measurements were carried out using a Agilent 1100 series HPLC (Germany) consisting of a gradient pump, a YMC Pack ODS-AM (5µm, 250 mm × 4,6 mm ID) end-capped column coupled with an UV-Vis detector and computer. The eluent was acetonitrile-water mixture (%50 (v/v)) in Britton-Robinson buffer (pH 2.5) at a flow rate of 1 mL·min^{1}. All separation was carried out at 25°C. The detection was performed at 250 nm.

3.3. Preparation of solutions

The voltammetric behaviour of sulfamethoxazole was investigated at Britton-Robinson (BR) buffer solutions prepared in 15, 30, 40 and 50 % (v/v) acetonitrile-water binary mixture at pH value from 1.5 to 7.0. A stock BR buffer solution composed of mixture of boric, acetic and phosphoric acids (each 0.04 M) were prepared and its pH values were adjusted by the addition of 1.0 M NaOH. pH meter was calibrated in the acetonitrile-water solution with different hydro-organic mixtures potassium hydrogenphthalate buffers (0.05 mol·kg^{-1}). A standard stock solution of sulfamethoxazole (10 mM) was obtained by dissolving 0.0253 g of sulfamethoxazole in a 10 mL volumetric flask for each acetonitrile-water mixture.

Two pharmaceutical tablets having sulfamethoxazole were weighed and powdered homogeneously in a mortar. A quantity of the powder, equivalent to one tablet, was dissolved in 50% acetonitrile-water mixture. The solution was sonicated for 10 min and filtered An aliquot of appropriate volume of stock solution was transferred into 250 mL volumetric flask and volume was completed with BR buffer (50% acetonitrile-water mixture, pH 2.5). The syrup was transferred to a 100 mL flask and volume was adjusted to the same pH value with buffer solutions. The samples were then spiked with appropriate amount of sulfamethoxazole for experiments.

3.4. Preparation of MIP and NIP electrodes

A Noki pencil model 2000 (Japan) was used as a holder for graphite leads (Tombo, HB, 0.5 mm diameter, Japan). The PGE was prepared by cutting the leads into 3 cm long sticks. Electrical contact with the lead was obtained by soldering a metal wire to the metallic part. PGEs were washed with acetonitrile to remove the impurity and dried at room temperature before use. Then, PGE was immersed the polymerization solution.

The MIP was obtained by electrodeposition on the surface of the PGE using cyclic voltammetry in the potential range between -0.60 and +1.40 V during five cycles (scan rate: 100 mV/s). The polymerization solution includes 0.1 M tetrabutylammonium perchlorate (TBAP), 0.1 M pyrrole and 0.005 M sulfamethoxazole in acetonitrile. After the electropolymerization process, the embedded sulfamethoxazole were then extracted to give a surface complimentary in shape and functionality to the original template sulfamethoxazole. The imprinted polymer was overoxidized in 0.1 M NaOH solution. The overoxidation was performed by scanning the potential repeatedly in a range of +0.80 to +1.20 V at a scan rate of 50 mV/s. We determined the sulfamethoxazole in this solution using DPV by uncovered PGE. Figure 11 shows that the template extracted from the polymer transferred in to this solution. A control electrode (non-imprinted overoxidized polypyrrole modified electrode, OPPY-NIP) was prepared in every case under the same experimental conditions but without adding the sulfamethoxazole, to check the reliability of the measurements.

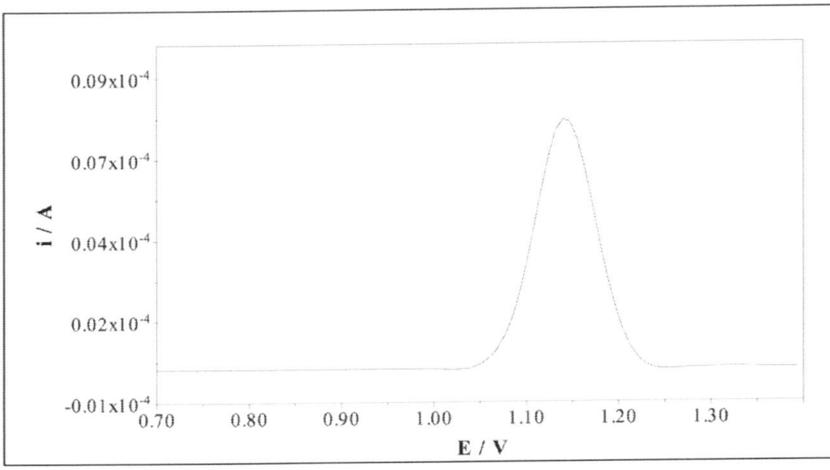

Figure 11. Differential pulse voltammogram of template by PGE in a desorption solution.

3.5. Electroanalytical measurements

Voltammetric measurements were carried out in three-electrode cell, at Britton-Robinson (BR) buffer solutions prepared in 15, 30, 40 and 50% (v/v) acetonitrile-water binary mixture from pH 1.5 to 7.0. Before the measurements, electrolytic solutions were purged with nitrogen for 5 min. Current measurements were performed using differential pulse voltammetry in a potential range between 0.00 and 1.40 V at a scan rate of 15 mV s^{-1}, modulation amplitude of 50 mV and step potential of 8 mV. All electroanalytical measurements were made at room temperature.

4. CONCLUSIONS

In this study, sulfamethoxazole imprinted electrodes formed by the cyclic voltammetric deposition of polypyrrole film on pencil graphite electrode in the presence of sulfamethoxazole have been fabricated. The molecularly imprinted electrode has been successfully applied as a sensor for fast and accurate determination of sulfamethoxazole in standards and pharmaceutical samples. The molecularly imprinted film exhibited a high sensitivity to sulfamethoxazole in 50% (v/v) acetonitrile-water mixture at pH 2.5. A linear relationship between sulfamethoxazole concentration and current response was obtained with excellent reproducibility of the current and a low detection limit of $3.59.10^{-7}$. The result of limit detection is lower than that reported in the literature using HPLC [6] and capillary electrophoresis [8] methods. Additionally, compared with results on application of molecularly imprinted polymer reported by other authors [16, 19, 21], the detection limit determined using MIP-based sensor indicate no significant difference.

The MIP based sensor was used for determination of sulfamethoxazole in commercial pharmaceutical samples. The recoveries in tablet and syrup were found as 96.0 and 87.4 % with the relative standard deviations of 0.90 and 0.85%, respectively. The sulfamethoxazole concentrations determined using this method is in good agreement with declared values. The proposed low cost chemical sensor could find application in the measurement of sulfamethoxazole level in clinical samples as well as in pharmaceutical industry.

ACKNOWLEDGMENTS

Financial support of Süleyman Demirel University (Projects No: 1202-D-05) and Anadolu University (Project No: 031064) Research Founds is gratefully acknowledged.

REFERENCES

1. Zheng, N.; Li, Y.; Wen, M. Sulfamethoxazole-imprinted polymer for selective determination of sulfamethoxazole in tablets. J. Chromatogr. A **2004**, 1003, 179–182.
2. Commission of the European Community. The rules governing medical products in the European Community VI.; CEC: Brussels; October; 1991.
3. Reeves, V.B. Confirmation of multiple sulfonamide residues in bovine milk by gas-positive chemical ionization mass spectrometry. J. Chromatogr. B **1999**, 723, 127–137.
4. Klimes, J.; Mokry, M. High performance liquid chromatography analysis of selected sulfonamides in plasma. Pharmazie **1997**, 52, 448–450.
5. Stoev, G.; Michailova, A. Quantitative determination of sulfonamide residue in foods of animal origin by high-performance liquid chromatography with fluorescence. J. Chromatogr. A **2000**, 871, 37–42.

6. Akay, C.; Özkan, S.A. Simultaneous LC determination of trimethoprim and sulphamethoxazole in pharmaceutical formulations. J. Pharm. Biomed. Anal. **2002**, 30, 1207–1213.

7. Zayas-Blanco, F.; Garcia-Falcon, M.S.; Simal-Gandara, J. Determination of sulfamethazine in milk by solid phase extraction and liquid chromatographic separation with ultraviolet detection. Food Control **2004**, 15, 375–378.

8. Berzas Nevado, J.J.; Castaneda Penalvo, G.; Guzman Bernardo, F.J. Determination of sulfametoxazole, sulfadiazine and associated compounds in pharmaceutical by capillary zone electrophoresis. J. Chromatogr. A **2001**, 918, 205–210.

9. Fuh, M.S.; Chu, S. Quantitative determination of sulfonamide in meat by solid-phase extraction and capillary electrophoresis. Anal. Chim. Acta **2003**, 499, 215–221.

10. Rao, T.N.; Sarada, B.V.; Tryk, D.A.; Fujishima, A. Electroanalytical study of sulfa drugs at diamond electrodes and their determination by HPLC with amperometric detection. J. Electroanal. Chem. **2000**, 491, 175–181.

11. Msagati, T.A.M.; Ngila, J.C. Voltammetric detection of sulfonamides at a poly(3-methylthiophene) electrode. Talanta **2002**, 58, 605–610.

12. Preechaworapun, A.; Chuanuwatanakul, S.; Einaga, Y.; Grudpan, K.; Motomizu, S.; Chailapakul, O. Electroanalysis of sulfonamides by injection system/high-performance liquid chromatography coupled with amperometric detection using boron-doped diamond electrode. Talanta **2005**, 68, 1726–1731.

13. Haupt, K. Molecularly imprinted polymers in analytical chemistry. Analyst **2001**, 126, 747–756.

14. Piletsky, S.A.; Piletskaya, E.V.; Sergeyeva, T.A.; Panasyuk, T.L.; El'skaya, A.V. Molecularly imprinted self-assembled films with specificity to cholesterol. Sens. Actuat. B **1999**, 60, 216–220.

15. Malitesta, C.; Losito, I.; Zambonin, P.G. Molecularly imprinted electrosynthesized polymers: New materials for biomimetic sensors. Anal. Chem. **1999**, 71, 1366–1370.

16. Gong, J.; Gong, Fu.; Zeng, G.; Shen, G.; Yu, R. A novel electrosynthesized polymer applied to molecular imprinting technology. Talanta **2003**, 61, 447–453.

17. Ramanavicius, A.; Ramanaviciene, A.; Malinauskas, A. Electrochemical sensors based on conducting polymer-polypyrrole. Electrochim. Acta **2006**, 6025–6037.

18. Panasyuk, T.L.; Mirsky, V.M.; Piletsky, S.A.; Wolfbeis, O.S. Electropolymerized molecularly imprinted polymers as receptor layers in capacitive chemical sensors. Anal. Chem. **1999**, 71, 4609–4613.

19. Gomez-Caballero, A.; Goicolea, M.A.; Barrio, R.J. Paracetamol voltammetric microsensors based on electrocopolymerized-molecularly imprinted film modified carbon fiber microelectrodes. Analyst **2005**, 130, 1012–1018.

20. Deore, B.; Chen, Z.; Nagaoka, T. Potential-induced enantioselective uptake of amino acid into molecularly imprinted overoxidized polypyrrole. Anal. Chem. **2000**, 72, 3989–3994.

21. Ebarvia, B.S.; Cabanilla, S.; Sevilla, F. Biomimetic properties and surface studies of a piezoelectric caffeine sensor based on electrosynthesized polypyrrole. Talanta **2005**, 66, 145–152.

22. Liang, H.; Ling, T.; Rick, J.F.; Chou, T. Molecularly imprinted electrochemical sensor able to enantroselectivly regonize D and L-tyrosine. Anal. Chim. Acta **2005**, 542, 83–89.

23. Ramanaviciene, A.; Finkelsteinas, A.; Ramanavicius, A. Basic electrochemistry meets nanotechnology: Elelectrochemical preparation of artificial receptors based on nanostructured conducting polymer, polypyrrole. J. Chem. Educ. **2006**, 83, 1212–1214.

24. Özcan, L.; Şahin, Y. Determination of paracetamol based on electropolymerized-molecularly imprinted polypyrrole modified pencil graphite electrode. Sens. Actuat. B **2007**, 127, 362–369.

25. Özcan, L.; Şahin, M.; Şahin, Y. Electrochemical preparation of a molecularly imprinted polypyrrole-modified pencil graphite electrode for determination of ascorbic acid. Sensors **2008**, 8, 5792–5805.

26. Majidi, M.R.; Jouyban, A.; Asadpour-Zeynali, K. Voltammetric behavior and determination of isoniazid in pharmaceutical by using overoxidized polypyrrole glassy carbon modified electrode. J. Electroanal. Chem. **2006**, 589, 32–37.

27. Ramanaviciene, A.; Ramanavicius, A. Molecularly imprinted polypyrrole-based synthetic receptor for direct detection of bovine leukemia virus glycoproteins. Biosens. Bioelectron. **2004**, 20, 1076–1082.

28. Mostany, J.; Scharifker, B.R. Impedance spectroscopy of undoped, doped and overoxidized poypyrrole films. Synth. Met. **1997**, 87, 179–185.

29. Ersöz, A.; Gavalas, V.G.; Bachas, L.G. Potentiometric behavior of electrodes based on overoxidized polypyrrole films. Anal. Bioanal.Chem. **2002**, 372, 786–790.

30. Tamer, U.; Ertaş, N.; Udum, Y.A.; Şahin, Y.; Pekmez, K.; Yıldız, A. Electrochemically controlled solid-phase microextraction (EC-SPME) based on overoxidized sulfonated polypyrrole. Talanta **2005**, 67, 245–251.

31. Şahin, Y.; Ercan, B.; Şahin, M. In situ electrochemical solid-phase extraction of anions and cations using polypyrrole and overoxidized sulfonated polypyrrole. Talanta **2008**, 75, 369–375.

32. Şahin, M.; Şahin, Y.; Özcan, A. Ion chromatography-potentiometric detection of inorganic anions and cations using polypyrrole and overoxidized polypyrrole electrode. Sens. Actuat. B **2008**, 133, 5–14.

33. Özcan, L.; Şahin, Y.; Türk, H. Non-enzymatic glucose biosensor based on overoxidized polypyrrole nanofiber electrode modified with cobalt (II) phthalocyanine tetrasulfonate. Biosens. Bioelectron. **2008**, 24, 512–517.

34. Hsueh, C.; Brajter-Toth, A. Electrochemical preparation and analytical applications of ultrathin overoxidized polypyrrole films. Anal.Chem. **1994**, 66, 2458–2464.

Chapter 11

High Sensitivity Electrochemical Cholesterol Sensor Utilizing a Vertically Aligned Carbon Nanotube Electrode with Electropolymerized Enzyme Immobilization

Anurat Wisitsoraat *, **Chanpen Karuwan, Krongkamol Wong-ek, Ditsayut Phokharatkul, Pornpimol Sritongkham and Adisorn Tuantranont**

Nanoelectronics and MEMS Laboratory, National Electronics and Computer Technology Center, 112 Thailand Science Park, Pahol Yothin Rd., Klong Luang, Pathumthani 12120, Thailand

ABSTRACT

In this report, a new cholesterol sensor is developed based on a vertically aligned CNT electrode with two-step electrochemical polymerized enzyme immobilization. Vertically aligned CNTs are selectively grown on a 1 mm^2 window of gold coated SiO_2/Si substrate by thermal chemical vapor deposition (CVD) with gravity effect and water-assisted etching. CNTs are then simultaneously functionalized and enzyme immobilized by electrochemical polymerization of polyaniline and cholesterol enzymes. Subsequently, ineffective enzymes are removed and new enzymes are electrochemically recharged. Scanning electron microscopic characterization indicates polymer-enzyme nanoparticle coating on CNT surface. Cyclic voltammogram (CV) measurements in cholesterol solution show the oxidation and reduction peaks centered around 450 and −220 mV, respectively. An approximately linear relationship between the cholesterol concentration and the response current could be observed in the concentration range of 50–300 mg/dl with a sensitivity of approximately 0.22 μA/mg·dl^{-1}, which is considerably higher compared to previously reported CNT bioprobe. In addition, good specificity toward glucose, uric acid acetaminophen and ascorbic acid have been obtained. Moreover, sensors have satisfactory stability, repeatability and life time. Therefore, the electropolymerized CNT bioprobe is promising for cholesterol detection in normal cholesterol concentration in human blood.

Keywords Vertically aligned varbon nanotubes; cholesterol; polyaniline; electropolymerization; enzyme based biosensor

1. INTRODUCTION

Cholesterol is an essential lipid for human body. The desired total plasma cholesterol for an individual is less than 5.2 mM (200 mg/dL) and it poses a potential health threat when the level is greater than 6.2 mM (240 mg/dL). Excessive plasma cholesterol causes poor cardiovascular conditions, such as atherosclerosis and hypertension, which can lead to coronary heart disease, myocardial and cerebral infarction. The total cholesterol in all lipoprotein fractions is normally determined using a colorimetric assay specific for cholesterol [1]. In the process, cholesterol esters are acted upon by cholesterol esterase (ChEs) to release free cholesterol and then cholesterol oxidase (ChOx) catalyzes the reaction to generate H_2O_2, which is reacted to yield quinoneimine dye. The absorbance of the dye is proportional to H_2O_2, hence the cholesterol concentration. The disadvantages of this technique include instability of color reagent, variable reactivity of ester, corrosive nature of reagents, poor specificity and unreusable reagents.

Electrochemical detection techniques emulate the colorimetric assay by immobilizing ChEs and ChOx onto electrode surfaces to liberate cholesterol and subsequently generate H_2O_2, which is measured amperometrically [2]. Its advantages include rapid analysis, reusability, thermal stability, linearity and good specificity. However, electrochemical biosensors still need further development to improve performances in term of sensitivity, reproducibility and life time. Several approaches including functionalization of the working electrode surface by polymer or nanomaterials have been proposed to enhance the performance of electrochemical biosensors [3,4].

Carbon nanotubes (CNTs) have been shown to promote electron transfer reactions for the redox of important biomolecules [5-8]. In addition, vertical alignment of the CNT nanoeletrodes is preferred over other schemes because the open end of a CNT is expected to show faster electron transfer rate (ETR). This characteristic along with their high surface area and conformal compatibility with biomolecules makes CNTs particularly attractive for high-sensitivity biochemical sensing systems. However, the performance of CNTs based electrochemical biosensor is also critically dependant on effectiveness of enzyme immobilization on CNT structure. Conventional approaches for enzyme immobilization on amperometric bioprobes include direct adsorption [9] and entrapment in polymeric films [10]. Physical adsorption technique suffers from protein desorption due to changes in temperature, pH and ionic strength while the enzyme entrapment in the lattice of a polymer matrix or membrane provides relatively better enzyme retention.

Recently, vertically aligned CNT electrode with enzyme entrapment in polyvinyl alcohol (PVA) by spin coating has been reported [8]. Nevertheless, the disadvantages of the reported technique are low enzyme entrapment efficiency and life-time due to PVA's low physical entrapment capacity, low conductivity and partial water solubility. Electrochemical enzyme entrapment with conducting polymer should offer much higher enzyme retention capacity and better electron transfer to CNTs [12-16]. There have been some reports on biosensors based on CNTs-enzyme-polymer composites prepared by electrochemical polymerization [17-22]. However, there have been few reports that perform enzyme immobilization by electrochemical polymerization on vertically aligned CNTs electrode, which can offer significantly improved sensor's performance and reliability. In this work, cholesterol bioprobe is developed based on vertically aligned CNTs with enzyme immobilization in polyaniline (PANI) using two-step electrochemical process.

2. EXPERIMENTAL SECTION

The structure for electrode fabrication is shown in Figure 1. First, SiO_2 (400 nm), Cr (50 nm) and Au (500 nm) were successively sputtered on <100> Si substrates. Next, aluminium oxide (10 nm) and stainless steel (SS) catalyst (5 nm) were sequentially sputtered to prepare for CNT synthesis. Titanium dioxide was then sputtered via shadow masking on the gold layer over a defined region, which excludes active sensing area (1 mm^2) and electrical contact region. The aluminium oxide and titanium dioxide layers were deposited by reactive sputtering at a pressure of 3×10^{-3} mbar of 1:5 Ar/O_2 gas mixtures while other metallic layers were deposited in Ar gas at the same pressure.

Vertically aligned carbon nanotubes were grown by thermal chemical vapor deposition (CVD) with gravity effect and water-assisted etching. The catalyst layers on substrates were placed upside down along gravitational field on an alumina carrier in a horizontal furnace thermal CVD system. The CNT synthesis was conducted at atmospheric pressure and growth temperature of 700 °C. During CNT growth, acetylene was flown for 1.5 minute and hydrogen to acetylene volume flow ratio was 4.3:1 with hydrogen flow of 1,935 sccm and acetylene flow of 450 sccm. In the course of CNT growth, in-situ-water-assisted etching was employed to remove undesired amorphous carbon formation from random acetylene decomposition. In water etching process, 300 ppm of water vapor was introduced by water bubbling through Ar gas for 3 minutes while acetylene gas was turned off. CNTs growth and water-assisted etching were repeatedly performed for three cycles and the total time was 13.5 minutes. CNTs' functionalization and enzyme immobilization in PANI matrix were then conducted as per the following protocol.

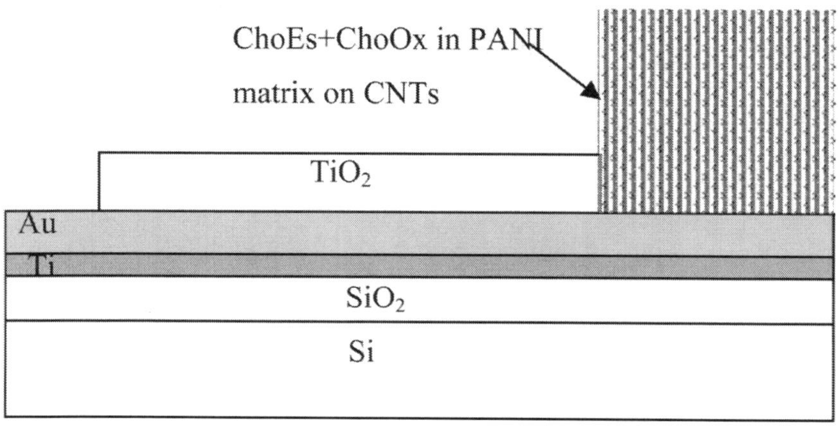

Figure 1. Structure of the CNT based cholesterol bioprobe.

Reagents for biosensing

Analytical-grade bio-reagents used in this study: cholesterol (95%), cholesterol oxidase (ChOx, Pseudomonas fluorescens 25 U/mg), cholesterol esterase (ChEs, Pseudomonas fluorescens 100 U/mg), horseradish peroxidase (HRP, 5,000 U), potassium ferrocyanide, trehalose, Triton X-100 were purchased from Sigma-Aldrich. Aniline monomer and phosphate buffer saline (PBS) solution was purchased from Fluka (USA).

Surface functionalization and enzyme immobilization

Functionalization of the CNT based electrode was done by electrochemical polymerization on the active zone of the working electrode in a solution of 0.2 M aniline monomer. Prior to aniline polymerization, CNT was refluxed in 1 M nitric acid for 2 hours at room temperature to remove catalyst and clean CNT surface. The aniline solution was prepared by dissolving 4.8 mL of aniline monomer and 1 mL of HCl in 100 mL of PBS solution (pH 7.0) with constant stirring for 10 minutes at room temperature. The final pH of aniline solution is around 2.5. Electrochemical polymerization was performed at ~ 20 °C by a standard three electrode electrochemical system (Metrohm electrochemical workstation with BAS C3 voltammetric cell). The working electrode (WE) was the CNT based electrode while the counter electrode (CE) was a Pt wire and all the potentials were controlled and measured as referred to an Ag|AgCl|KCl (sat.) reference electrode (RE). During polymerization, the CNT electrode was biased at a constant voltage of 0.6 V with respect to RE for 30 minutes. The polyaniline matrix formed on CNTs introduced NH group on CNT surface, which greatly helped immobilizing various biological analytes, particularly proteins [12,13].

Enzyme immobilization was then followed by a two-step electrochemical procedure. The first electrochemical process was conducted in the mixture of 2 mL of cholesterol enzyme and 2 mL of 0.2 M aniline monomer. The pH of the mixture was measured to be about 3. The cholesterol enzyme was prepared by dissolving 1.5 mg ChEs, 2.0 mg ChOx, 1.8 mg HRP, 8.0 mg potassium ferrocyanide (redox mediator) and 6.0 mg trehalose (enzyme stabilizer) in 0.3 mL PBS solution. The PANI-CNT electrode was biased at a constant voltage of 0.6 V with respect to RE for 30 minutes. Due to the acidic condition during aniline polymerization, cholesterol enzymes were inactive or even denatured. Thus, ineffective enzymes were removed by etching in stirred 6 M HCl solution for 2 hours, leaving cavities in and on PANI matrix. The PANI layer was then reduced by biasing at constant voltage of −0.2 V for 20 minutes in PBS solution to remove any residual anions from acidic etching. New enzymes were electrochemically recharged into PANI matrix by biasing the electrode in cholesterol enzyme in PBS solution (pH 6.0) at a constant voltage of 0.6 V with respect to RE for 30 minutes. After immobilization, the sensor chip was stored in a refrigerator for 24 h and thoroughly rinsed in PBS before running each test.

Electrochemical measurements

Electrochemical measurement of cholesterol was performed by the same electrochemical system. Due to cholesterol's water insolubility, the stock cholesterol solution was prepared by dissolving 40 mg cholesterol in 1 mL Triton X-100 and 9 mL 0.1M PBS solution (pH 7.0) and stirring with a magnetic spin bar just before the measurement to obtain a homogeneous solution. Cholesterol solution with different concentration was then prepared for testing by proper dilution of the stock solution. Cholesterol bioprobe was immersed in 0.1 M PBS solution and cyclic voltammograms (CV) were run at several scan rates for different concentrations of cholesterol. The voltage window for CV was from −1.0 to +1.0 V because the redox peaks appeared within this potential window.

3. RESULTS AND DISCUSSION

The surface morphology of the sensing area of CNT based biochip was examined using scanning electron microscopy (SEM). Scanning electron microscopic images of vertically aligned CNTs before and after polyaniline-enzyme immobilization,is shown in Figures 2(a) and (b), respectively. The cross section seen in the images confirms that CNTs are indeed vertically aligned, with a vertical height of ~50 μm. In addition, it can be seen that vertically aligned CNTs are uniformly coated by polymer-enzyme nanoparticles on CNT surface. Figure 3(a) shows high magnification top view SEM image of CNT electrodes after electrochemical polyaniline-enzyme coating. It can be seen the size of polyaniline-enzyme nanoparticles is in the range between 20 and 50 nm. In addition, transmission electron microscopic (TEM) characterization as shown in Figure 3(b) verifies that CNTs are multiwalls with diameters and numbers of walls in the range of 10–25 nm and 8–15 respectively.

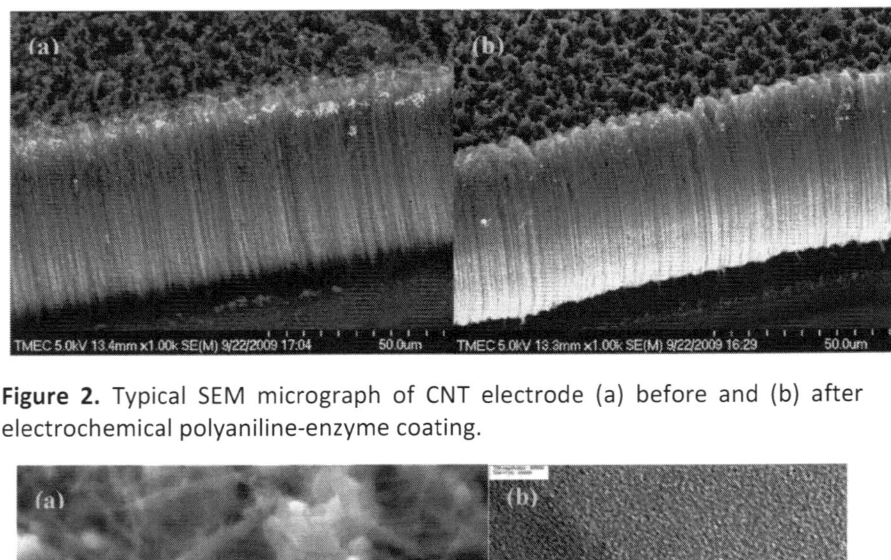

Figure 2. Typical SEM micrograph of CNT electrode (a) before and (b) after electrochemical polyaniline-enzyme coating.

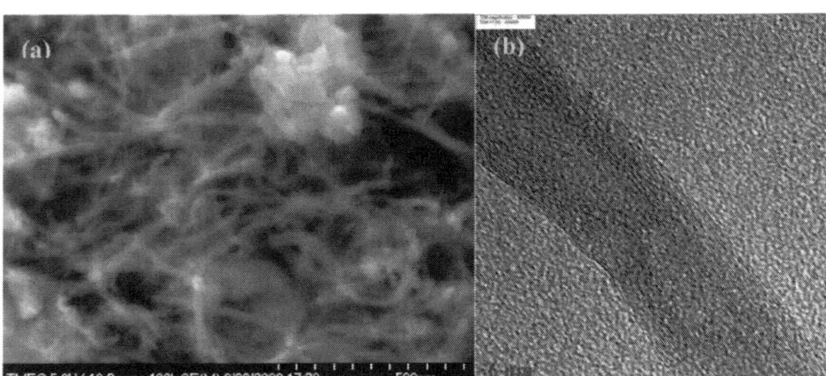

Figure 3. High magnification (a) top view SEM image of CNT electrodes after electrochemical polyaniline-enzyme coating and (b) TEM image of CNT extracted from the synthesized vertically aligned CNTs.

The chemical property of the CNT bioprobe was characterized by Fourier Transform Infrared Spectroscopy (FTIR) using a Perkin Elmer System—2000R FTIR spectrophotometer. The FTIR spectrum of the CNT bioprobe is shown in Figure 4. The observed functional groups confirm successful polyaniline formation and cholesterol enzyme immobilization. The PANI, ChOx, and ChEs exhibit common characteristic bands at around 2,852, 1,565, and 1,300 cm^{-1}, which are the C–H bending vibration out of the plane of the para-disubstituted benzene rings, the C–C stretching band in benzenoid rings, the C–N stretching mode. In addition, PANI have distinctive peaks at 1,143 and 796 cm^{-1}, the vibration band of the anion (HCl–PAni), and N–H vibration band of aniline, respectively. Likewise, ChOx and ChEs have distinct absorption peaks at approximately 1,424, 2,867 and 3,524 cm^{-1}, which correspond to the CH_3, C–CH_3, OH vibration bands.

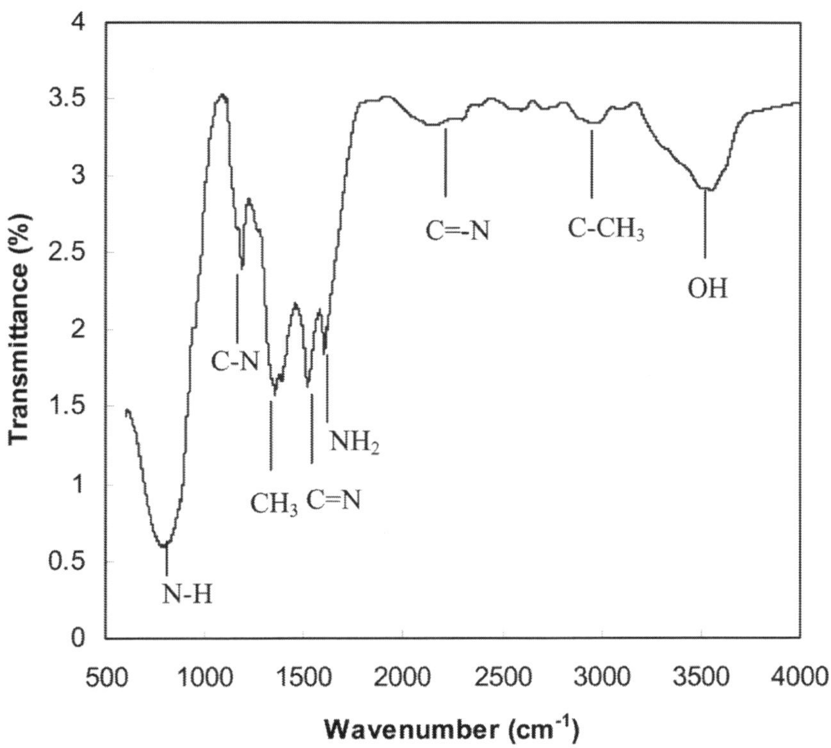

Figure 4. Typical FT-IR spectrum of the CNT electrode after electrochemical polyaniline-enzyme coating.

CV curves for gold based and CNT based cholesterol bioprobe at a cholesterol concentration of 300 mg·dL^{-1} and 100 mV scan rate are shown in Figure 5. The gold based bioprobe was prepared by the same two-step electrochemical procedure as that was done for CNT based bioprobe. It is evident that CNT based bioprobe show much higher current sensitivity than gold based cholesterol bioprobe. The result confirms the enhancement of electrochemical detection of cholesterol by vertically aligned CNT structures. However, this result is in contrast to Lee's report [22] that Pd sputtered screen-printed carbon electrodes (SPCEs) exhibited higher electrochemical sensitivity than CNT modified SPCEs. The contradiction can be explained from the fact that the substrate surfaces for sputtered thin film coating are significantly different.

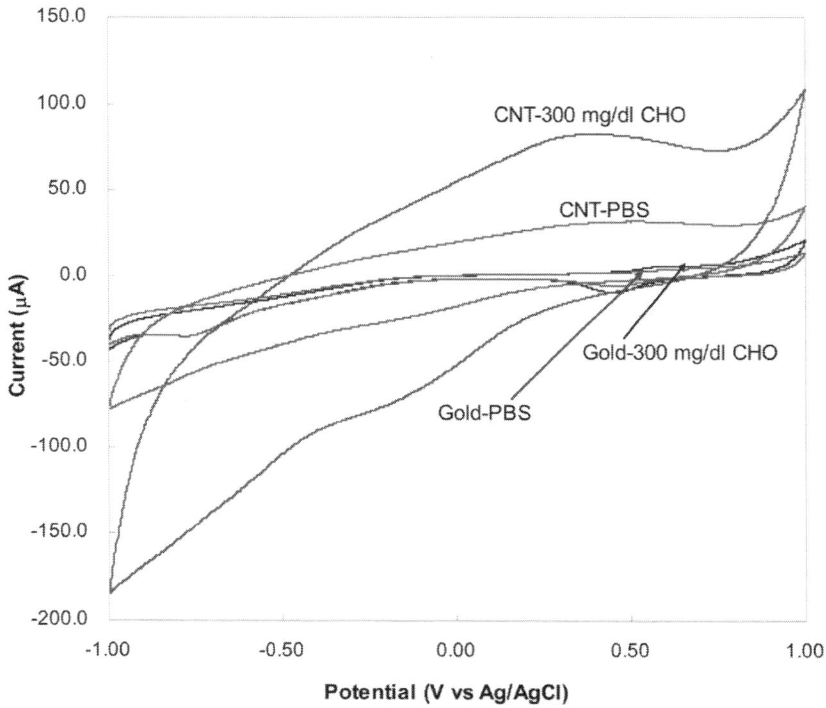

Figure 5. Cyclic voltammograms for gold and CNT based cholesterol bioprobe at 300 mg/dL concentration and 100 mV scan rates coating.

In Lee's work, screen-printed carbon electrodes (SPCEs) were used but standard SiO_2/Si substrates were utilized in this work. The surfaces of SPCEs were very rough and pitted while SiO_2/Si substrates were very smooth. The Pd sputtering caused better enhancement than CNTs modification on SPCEs because thin Pd sputtering could directly increase electron transfer rate without losing surface area whereas CNTs modification would reduce electrode surface area as Nafion/CNTs filled into pores of SPCE surface. In our work, vertically aligned CNTs that were directly grown on gold sputtered SiO_2/Si electrode significantly increased specific surface area and electron transfer rate.

From Figure 5, the anodic and the cathodic peaks at scan rate of 0.1 V/s are centered on 397 and 526 mV, respectively. This could be ascribed to the electrocatalytic reduction of H_2O_2 by horseradish peroxidase following cholesterol release and decomposition by ChEs and ChOx, respectively [11], as illustrated in the following set of reactions:

$$\text{cholesterol ester} \rightarrow \text{fatty acid} + \text{cholesterol (cholesterol esterase)} \quad (1)$$
$$\text{cholesterol} \rightarrow \text{cholest-4-ene-3 one} + H_2O_2 \text{ (cholesterol oxidase)} \quad (2)$$
$$H_2O_2 + 2[Fe(CN)_6]^{4-} + 2H^+ \longrightarrow 2[Fe(CN)_6]^{3-} + 2H_2O \text{ (horseradish peroxidase)} \quad (3)$$

CV curves for cholesterol bioprobe at a cholesterol concentration of 300 $mg.dl^{-1}$ with different scan rates are shown in Figure 6. It can be noticed from the given set of CV curves that magnitude of the anodic and cathodic peaks are different for different scan rates. The anodic peaks shift from 490 mV to 540 mV as the scan rate increases from 10 mV/s to 300 mV/s while the cathodic peaks change from −250 mV to −333 mV. Moreover, it can be seen that the redox reaction is not reversible since the peak current ratio between anodic and cathodic (i_{pa}/i_{pc}) is not equal to one for all scan rates. This implies that the electron transfer rate in the chemical reactions, which are coupled to the redox process or to the adsorption of either reactants or products that might occur during the scanning process, is slow relative to the voltage scan rate. Furthermore, the oxidation peak amplitude is found to increase linearly with the square root of scan rate as shown in the inset of Figure 6, indicating that the current is limited by semi-infinite diffusion of cholesterol on the CNTs electrode.

Five independent measurements were made to assess repeatability and stability of the sensors. The error bars shown in the inset of Figure 6 demonstrate that the sensor has good repeatability and stability of less than 10% under different scan rates.

Figure 7 represents the set of waveforms obtained for different concentrations of cholesterol at a specific scan rate (100 mV/s). The difference in the peak current would be the detection capacity for the sensor in a cholesterol concentration at this working electrode surface of 1 mm^2 active area. It can be seen that the reduction and oxidation peaks become clearly pronounced as the cholesterol concentration increases from 100 to 300 mg/dL. The oxidation and reduction peaks centered around 450 and −250 mV, respectively. In addition, the peaks are more evident and located at higher voltage as scan rate increases at a fixed concentration.

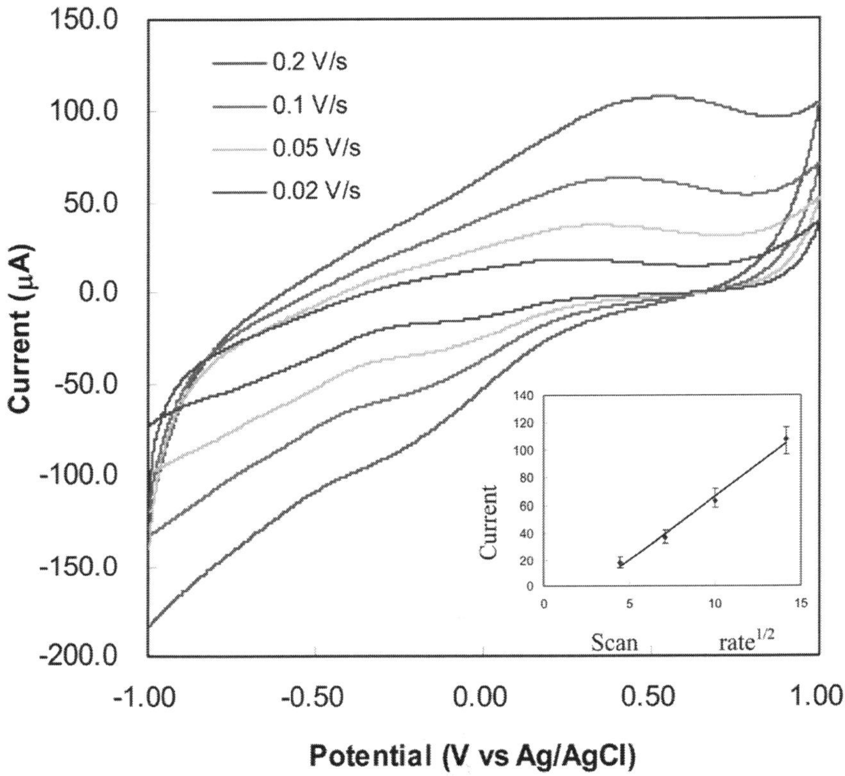

Figure 6. Cyclic voltammograms for cholesterol bioprobe at 300 mg/dL concentration with different scan rates. Inset: Peak current (oxidation peaks) as a function of the square root of scan rate. The error bars indicates the variation from five independent measurements.

The amperometric response (inset of Figure 7) is the amplitude of oxidation peaks above background from CV curves. An approximately linear relationship between the analyte concentration and the response current of the CNT based electrode can be observed between 50 and 300 mg/dL concentration with a sensitivity of approximately 0.22 $\mu A/mg{\cdot}dL^{-1}$, which is considerably higher compared to previously reported CNT bioprobe [8]. The electrochemical sensing performance of the electrode can be attributed to the combination of CNTs properties and efficient enzyme immobilization by two step electrochemical process. Since the response current appeared to be in linear relation with the cholesterol concentration, so the enzyme catalytic reaction of cholesterol is the first-order reaction. The error bars shown in the inset of Figure 7 demonstrate that the sensor has good repeatability and stability of less than 15% over the whole concentration range.

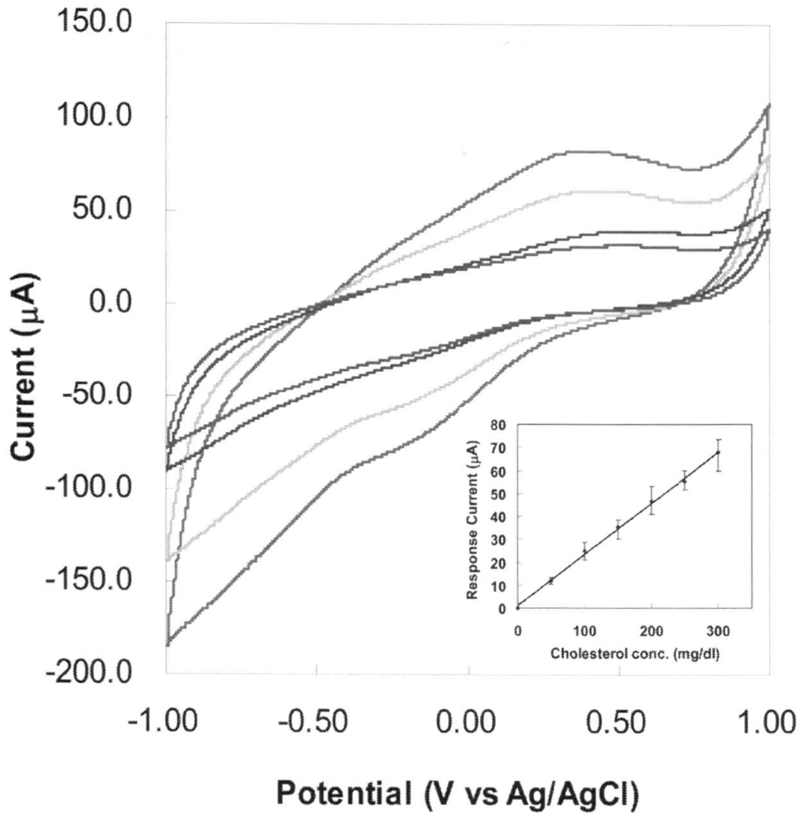

Figure 7. Cyclic voltammograms for cholesterol bioprobe at 100 mV/s scan rate with different cholesterol concentrations. Inset: Amperometric response for bioprobe (oxidation peaks) as a function of cholesterol concentration. The error bars indicates the variation from five independent measurements.

Specificity is one of the most critical issues for biosensors to be used in real environment. To evaluate specificity of the sensors, interferences by four common electroactive species including ascorbic acid, glucose, acetaminophen and uric acid were tested at their normal higher concentration level in serum.

Figure 8 demonstrates interference peaks of ascorbic acid (1 mg/dL), glucose (150 mg/dL), acetaminophen and uric acid (7 mg/dL) along with the peak of cholesterol (200 mg/dL). It can be seen that ascorbic acid and glucose give very low interference signals, while acetaminophen and uric acid produce considerable interference. Nevertheless, the interference from these analytes can be minimized by setting the working electrode potential at its peak value

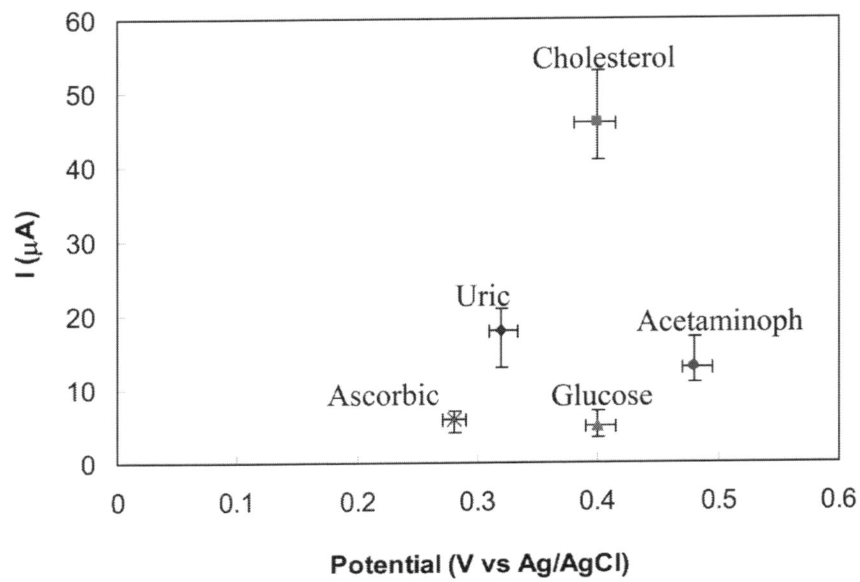

Figure 8. Oxidation peak current vs. corresponding peak potential of cholesterol (200 mg/dL), ascorbic acid (1 mg/dL), glucose (150 mg/dL), acetaminophen (5 mg/dL) and uric acid (7 mg/dL). The error bars indicates the variation for five independent measurements.

(~0.4 V) since the peaks of interfering analytes locate at different potentials except glucose which has similar peak location but its interference is very low. Finally, long-term stability of the sensor was assessed. The sensors were stored dry at 4 °C and tested every day. It was found that the sensitivity was dropped by 15% after one month due to natural enzyme degradation. Thus, the sensors have a satisfactory life time. Therefore, the present sensor is promising for clinical diagnostics after the calibration to be done with real blood samples.

4. CONCLUSIONS

In conclusion, electropolymerized vertically aligned carbon nanotubes have been developed to enhance electrochemical sensing of gold coated Si electrode. The fabricated sensors have been characterized for cholesterol detection in the concentration range between 0 and 300 mg/dL cholesterol concentration by standard cyclic voltammogram measurement. An almost linear relationship between the analyte concentration and the response current of the CNT based electrode could be observed between 100 and 300 mg/dL concentration with a high current sensitivity. In addition, good specificity toward glucose, uric acid acetaminophen and ascorbic acid have been obtained. Moreover, sensors have satisfactory stability, repeatability and life time. Thus, the electropolymerized CNT bioprobe can effectively be used for cholesterol detection in normal range of cholesterol concentration in human blood. In addition, this developed scheme has the potentiality to integrate an array of sensors for lab-on-a-chip systems due to its compatibility with the standard Si micro-fabrication technology.

REFERENCES

1. Allain, C.C.; Poon, L.S.; Chan, C.S. Enzymatic determination of total serum cholesterol. Clin. Chem. **1974**, 20, 470–475.
2. McCreery, R.L. Electroanalytical Chemistry, 1st ed.; Marcel Dekker: New York, NY, USA, 1991; pp. 221–228.
3. Bescombes, J.L.; Cosnier, S.; Labbe, P. Polyphenol oxidase-catechol an electroenzymatic model system for characterizing the performance of matrices for biosensors. Talanta **1996**, 43, 1615–1619.
4. Yeh, J.I.; Lazareck, A.; Kim, J.H.; Xu, J.; Du, S. Peptide nanowires for coordination and signal transduction of peroxidase biosensors to carbon nanotube electrode arrays. Biosens. Bioelectron. **2007**, 23, 568–574.
5. Ohba, T.; Kaneko, K. Internal surface area evaluation of carbon nanotube with GCMC simulation-assisted N_2 adsorption. J. Phys. Chem. B **2002**, 106, 7171–7176.
6. Li, G.; Liao, J.M.; Hu, G.Q.; Ma, N.Z.; Wu, P.J. Study of carbon nanotube modified biosensor for monitoring total cholesterol in blood. Biosens. Bioelectron. **2005**, 20, 2140–2144.
7. Guiseppi-Elie, A.; Lei, C.; Baughman, R.H. Direct electron transfer of glucose oxidase on carbon nanotubes. Nanotechnology **2002**, 13, 559–564.
8. Roy, S.; Vedala, H.; Choi, W.B. Vertically aligned carbon nanotube probes for monitoring blood cholesterol. Nanotechnology **2006**, 17, 14–18.
9. Ferri, T.; Poscia, A.; Santucci, R. Direct electrochemistry of membrane-entrapped horseradish peroxidase. Part I A voltammetric and spectroscopic study. Bioelectrochem. Bioenergy **1998**, 44, 177–181.

10. Su, X.; Zong, Y.; Richter, R.; Knoll, W. Enzyme immobilization on poly(ethylene-co-acrylic acid) films studied by quartz crystal microbalance with dissipation monitoring. J. Colloid Interface Sci. **2005**, 287, 35–42.

11. Kan, J.; Pan, X.; Chen, C. Polyaniline–uricase biosensor prepared with template process. Biosensors Bioelectron. **2004**, 19, 1635–1640.

12. Mu, S.; Xue, H. Bioelectrochemical characteristics of glucose oxidase immobilized in a polyaniline film. Sensors Actuators B **1996**, 3, 155–160.

13. Trojanowich, M.; Krawczyk, T. Krawczynsky vel Electrochemical biosensors based on enzymes immobilized in electropolymerized films. Microchim. Acta **1995**, 121, 167–181.

14. Losada, J.; Garcia Armada, M.P. A glucose amperometric sensor based on covalent immobilization of glucose oxidase in poly-2-aminoaniline film via chloranil on platinized platinum electrode. Electroanalysis **1997**, 9, 1416–1421.

15. Garjonyte, R.; Malinauskas, A. Amperometric glucose biosensors based on Prussian Blue—and polyaniline–glucose oxidase modified electrodes. Biosensors Bioelectron **2000**, 15, 445–451.

16. Wang, J.; Musameh, M. Carbon-nanotubes doped polypyrrole glucose biosensor. Analytica Chimica Acta **2005**, 539, 209–213.

17. Chen, G.; Shaffer, M.; Colbey, D.; Dixon, G.; Zhou, W.; Fray, D.; Windle, A. Carbon nanotube and polypyrrole composites coating and doping. Adv. Mater. **2000**, 12, 522–526.

18. Wei, Z.; Wan, M.; Lin, T.; Dai, L. Polyaniline nanotubes doped with sulfonated carbon nanotubes made via a self-assembly process. Adv. Mater. **2003**, 15, 136–139.

19. Wang, J.; Musameh, M.; Lin, Y. Solubilization of carbon nanotubes by nafion towards the preparation of amperometric biosensors. J. Am. Chem. Soc. **2003**, 125, 2408–2409.

20. Rubianes, M.D.; Rivas, G.A. Carbon nanotubes paste electrode. Electrochem. Commun. **2003**, 5, 689–694.

21. Luong, J.; Hrapovic, S.; Wang, D.; Bensebaa, F.; Simard, B. Solubilization of multiwall carbon nanotubes by 3-aminopropyltriethoxysilane towards the fabrication of electrochemical biosensors with promoted electron transfer. Electroanalysis **2004**, 16, 132–139.

22. Lee, C.-H.; Wang, S.-C.; Yuan, C.-J.; Wen, M.-F.; Chang, K.-S. Comparison of amperometric biosensors fabricated by palladium sputtering, palladium electrodeposition and Nafion/carbon nanotube casting on screen-printed carbon electrodes. Biosen. Bioelectron. **2007**, 22, 877–884.

Chapter 12

Fabrication of Poly(Aspartic Acid)-Nanogold Modified Electrode and Its Application for Simultaneous Determination of Dopamine, Ascorbic Acid, and Uric Acid

Juan Li, Xiaoli Zhang*

School of Chemistry and Chemical Engineering, Shandong University, Jinan, China

ABSTRACT

A nanostructured polymer film incorporated gold nanoparticles modified electrode was fabricated. The fabrication process involved a previous electropolymerization of aspartic acid and followed by the eletrodeposition of gold nanoparticles on the glassy carbon electrode. The resulting poly(aspartic acid)-nanogold modified electrode (PAAnano-Au/GCE) was characterized by scanning electron microscopy (SEM) and electrochemical impedance spectroscopy (EIS). A higher catalytic activity was obtained to electrocatalytic oxidation of dopamine (DA), ascorbic acid (AA) and uric acid (UA) due to the enhanced peak current and well-defined peak separations compared with three, bare GCE, PAA/GCE and nano-Au/GCE. Simultaneous determination of DA, AA, and UA were studied by voltammetry. The linear range of 5.0×10^{-7} - 1.0×10^{-4} M for DA, 5.0×10^{-6} - 2.0×10^{-3} M for AA and 5.0×10^{-6} - 1.0×10^{-3} M for UA was obtained. The detection limit was calculated for DA, AA and UA as being 6.5×10^{-8} M, 5.6×10^{-7} M and 3.0×10^{-7} M, respectively (S/N = 3). The practical application of the present modified electrode was demonstrated by the determination of DA, AA and UA in calf serum and fetal calf serum samples.

Keywords: Simultaneous Determination; Poly(Aspartic Acid)-Nanogold Modified Electrode; Dopamine; Ascorbic Acid; Uric Acid; Voltammetry

1. INTRODUCTION

Dopamine (DA) is an important neurotransmitter molecule and an indispensable intermediate species in human metabolism [1-3]. The loss of DA may cause some serious diseases, such as schizophrenia and Parkinson's disease [4]. Similarly, ascorbic acid (AA) functions as a redox cofactor and catalyst in a broad array of biochemical reactions and processes. AA is used for the prevention of scurvy and treatment of common cold, mental illness, cancer and AIDS [5]. Uric acid (UA) is produced by xanthine oxidase from xanthine and hypoxanthine, which are, in turn, produced from purine. UA is more toxic to tissues than either xanthine or hypoxanthine because, at high concentrations, it may cause such diseases as hyperuricaemia, gout, and the Lesch-Nyan disease [6]. Usually DA, AA and UA are coexisting in our body fluids. Therefore, it is very important to develop simple and rapid methods for their simultaneous determination in routine analysis.

Electrochemical detection of DA, AA and UA is a preferred method, because they are electro-chemically active. In addition, electrochemical method offers advantages such as simplicity, speed and sensitivity. A major problem in simultaneous determination is the resolution between DA and AA as well as DA and UA. At nearly all traditional bare electrodes, DA, AA and UA are oxidized at nearly same potential, resulting in an overlapping voltammetric response. To overcome the above problem, a number of studies as reported in the literature have been carried out to develop novel, modified electrodes for simultaneous determination of DA, AA and UA, such as silver doped poly(L-valine) modified GCE [7], zinc oxide composite film [8], palladium nanoparticle-loaded carbon nanofibers modified electrode [9], carbon ceramic electrode prepared by sol-gel technique [10], Pt-Au hybrid film modified electrode [11], carbon ionic liquid electrode [12], and ruthenium oxide modified electrode [13]. In recent years, electro-synthesized polymers modified electrode have been used to detect DA and/or DA, AA and UA, simultaneously, such as poly (vinyl alcohol) [14], poly(evans blue) [15], poly(oracet blue) [16], poly(chromotrope 2B) [17], poly(4-(2-pyridylazo) resorcinol) [18], poly(eriochrome black T) [19], poly(sulfonazo III) [20], poly(N,N'-dimethylaniline) [21], poly(3,4-ethylenedioxy) thiophene [22], poly(methylene blue) [23], poly(3,4-ethylenedioxy-thiopheneco-(5-amino-2-napthalenesulfonic acid)) [24] and poly(5-amino-2-mercapto-1,3,4-thia-diazole) [25] modified electrodes.

On the other hand, nanomaterials show higher electrocatalytic activity, as a kind of special sensing material of chemically modified electrodes. Particularly, the gold nanoparticles have attracted much attention in recent decades, owing to their excellent optical and electronic properties, as well as biocompatible ability. Very recently, we reported the electrodeposition preparation of nanogold modified glassy carbon electrode and its application to simultaneous voltammetric determination of dihydroxyhenzene isomers [26].

Composite electrodes have received great attention in the last decade due to their wide applications in analytical chemistry as chemical and biosensors [27]. Preparation of composite electrodes by incorporation of metal nanoparticles or polymers on the electrode surface can be extensively enhance the catalytic activity of the electrodes towards electrochemical behaviors of many biologically importance molecules [25,28].

The poly aspartic acid film has been used to fabricate modified electrode which can enhance the electron transfer rate and reduce the overpotential for the oxidation of AA [29]. Here we report for the first time a simultaneous voltammetric determination of DA, AA and UA using a poly aspartic acid incorporated nanogold particles modified glassy carbon electrode (PAA-nano-Au/GCE). It illustrated that the PAA-nano-Au/GCE not only can catalytically oxidize DA, AA and UA simultaneously, but separate their oxidation peaks distinctly. Compared with the existing reports about simultaneous determination of DA, AA and UA (shown in Table 1), the proposed method is much more convenient to prepare the modified electrode, the selectivity and stability is better.

2. EXPERIMENTAL METHOD

2.1. Chemicals

Dopamine was purchased from Sigma-Aldrich. Uric acid, ascorbic acid, $HAuCl_4$ and aspartic acid were obtained from Shanghai Chemical Reagents Co. Ltd. (Shanghai, China). All chemicals were of analytical reagent grade and used without further purification. Phosphate buffer solutions (PBS) with different pH values were prepared by mixing 0.10 M Na_2HPO_4 and 0.10 M NaH_2PO_4 and pH values were adjusted by addition of 1.0 M H_3PO_4 and/or NaOH solution. Freshly prepared solution of DA, AA, and UA were used in all experiments. Doubly distilled water was used throughout the experiments.

2.2. Apparatus

A CHI 800 electrochemical analyzer (Shanghai Chenhua Instrument Company, China) was used to perform electrode characterization and voltammetric measurements. A JEOL JSM-7600F Scanning Electron Microscope (SEM) was used to provide the information about the surface topography of electrodes. The Nyquist plots of the EIS measurement were recorded by the Potentiostat/Galvanostat Model 273A. A conventional three-electrode system was employed with a saturated calomel electrode (SCE) as the reference electrode, a platinum wire as the auxiliary electrode and a bare or modified glassy carbon electrode (GCE, 3 mm in diameter) as the working electrode. All potentials in the text were against SCE. The electrolyte solution was purged with highly purified nitrogen for at least 10 min before the experiment. The all experiments were sustained in the nitrogen atmosphere and carried out at room temperature.

2.3. Preparation of the Modified Electrode

The glassy carbon electrode was polished subsequently with 0.5 and 0.05 μm alumina slurry, then thoroughly rinsed with water and sonicated in ethanol and distilled water in turn. After being cleaned, the electrode was continuously CV scanned from –1.5 to 2.0 V at 100 $mV{\cdot}s^{-1}$ for 30 cycles in 10^{-3} M aspartic acid aqueous solution to form poly aspartic acid film on the electrode surface. Followed, this electrode was taken out and rinsed with water, then was immersed into 0.1 M KNO_3 containing 0.4 $g{\cdot}L^{-1}$ $HAuCl_4$, electrodeposition from $HAuCl_4$ was conducted for 60 s at –0.2 V. Finally, the resulting electrode, PAA-nano-Au/GCE, was activated by several successive cyclic voltammetric scanning from –0.5 to 1.0 V with 100 $mV{\cdot}s^{-1}$ in phosphate buffer solution (pH 6.5) until a steady voltammogram was obtained.

2.4. Analytical Procedure

The buffer solution was degassed with N_2 and kept under a N_2 blanket. After quantitative DA, AA and UA were introduced into the electrochemical cell, and then the voltammograms were recorded. The peak current was measured and recorded as a signal (I_p).

3. RESULTS AND DISCUSSION

3.1. Surface Morphology of Poly(Aspartic Acid)-Nanogold Film and Its Electrochemical Properties

Figure 1 displays the continuous cyclic voltammograms (CVs) of aspartic acid electropolymerization onto a GCE in 10^{-3} M aspartic acid aqueous solution over the potential range of –1.5 V to 2.0 V for 15 cycles at a scan rate

Table 1. Comparison of working range and detection limit of PAA-nano-Au/GCE with reported methods.

Linear range (μM)			Detection limit (μM)			Detection method	E_{pa} (V)			Reference
DA	AA	UA	DA	AA	UA		DA	AA	UA	
1 - 20	25 - 500	2.5 - 20	0.11	13	1.4	DPV	0.201	−0.021	0.339	[1]
6 - 960	15 - 240	50 - 800	0.7	1.4	4.5	CV	0.184	0.20	0.313	[8]
0.5 - 160	50 - 4000	2 - 200	0.2	15	0.7	DPV	0.402	0.158	0.550	[9]
0.5 - 50	5 - 1000	1 - 50	0.07	1.7	0.4	Amperometry	0.27	0.13	0.40	[10]
24 - 384	103 - 1650	21 - 336	24	103	21	DPV	0.35	0.20	0.54	[11]
2 - 70	10 - 250	2 - 50	1.4	7.6	0.6	DPV	0.17	−0.01	0.30	[14]
0.06 - 0.80 0.80 - 8.00	40.0 - 800.0	20.0 - 600.0	0.02	1.3	0.4	DPV	0.13	−0.35	0.225	[16]
20 - 200	150 - 1000	10 - 130	0.02	10	1.0	DPV	0.32	0.18	0.50	[19]
75 - 180	75 - 187	75 - 180	-	-	-	DPV	0.34	0.17	0.48	[30]
0.5 - 100	5 - 2000	5 - 1000	0.065	0.56	0.3	LSV	0.2	0.0	0.35	This work

Abbreviations: differential pulse voltammetry, DPV; linear sweep voltammetry, LSV; cyclic voltammetry, CV.

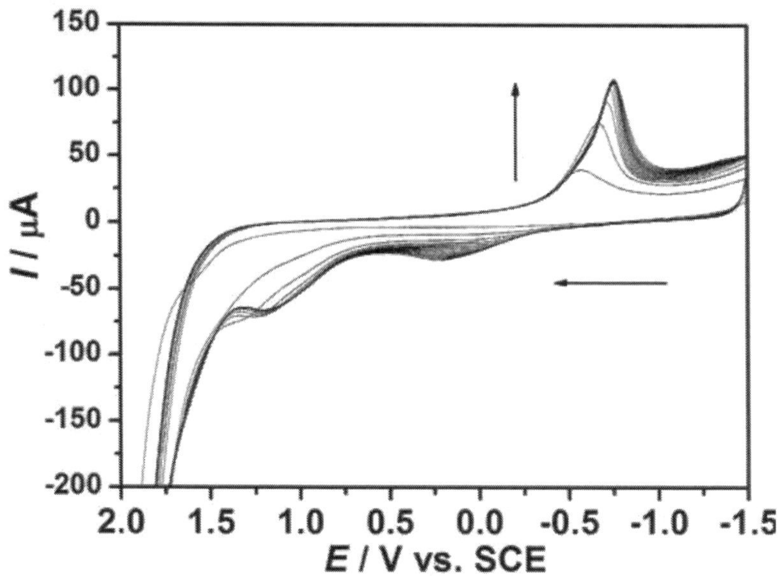

Figure 1. Cyclic voltammograms (CVs) of the poly aspartic acid film growth in the 10^{-3} M aspartic acid aqueous solution. Scan rate = 100 mV·s^{-1}.

of 100 mV·s^{-1}. As can be seen, all the anodic peak and cathodic peak currents enhanced gradually, indicating the formation and growth of an electroactive layer on the GCE surface. After the eighth cycle, the increase of these peaks current tended to be stable, implying that polymerization reach saturation. In addition, a uniform dark blue film presents on the surface of GCE, demonstrating the formation of poly aspartic acid film on GCE. As a primary amine, the electrochemical behavior of aspartic acid is in agreement with the literature [29,31,32]. We demonstrated that the potential range would affect the electropolymerization, only at high potential over 1.5 V, the oxidation of the amino group could turn into its corresponding cation radical, and these cation radicals can form carbon-nitrogen links at the carbon electrode surface. Accordingly, we propose this process illustrated in Scheme 1. The immobilized polymers through electropolymerization have many advantages in the detection of bio-molecules because of its selectivity, sensitivity and homogeneity in electrochemical deposition, strong adherence to electrode surface and chemical stability of the film.

The optimization of the PAA film thickness was also investigated. With the polymerizing cycles increasing, the electroresponse of DA, AA and UA increased at first. But when the polymerizing cycles is more than 15, the currents begin to fall. This could be explained that an increase in thickness of the film would prevent the electron transfer. Therefore, 15 cycles was used in the experiments.

Figure 2 shows the surface morphologies of bare electrode and modified electrode characterized by scanning electron microscopy (SEM). From the SEM images, It can be seen, the surface of bare GCE is smooth (Figure 2(a)). After separate gold nanoparticles were deposited on GCE, the surface presents spheroidal particles with well-distributed size, and particles' average diameter is about 50 nm (Figure 2(b)). Similarly, separate poly aspartic acid film on the GCE shows a more uniform surface topography with pycnotic nanostructure with round parvules (Figure 2(c)). After the aspartic acid was previously electropolymerizated on the GCE, the electrodeposition of gold nanoparticles were conducted for 60 s. The small snowflake-like particles could be observed at top level layer. These observations suggest that gold nanoparticles had been incorporated into the initially formed poly aspartic acid film (Figure 2(d)).

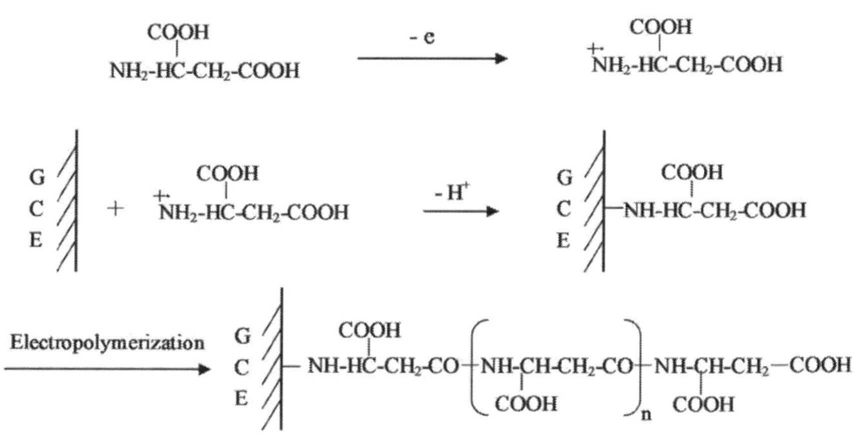

Scheme 1. The modification process of aspartic acid on glassy carbon electrode.

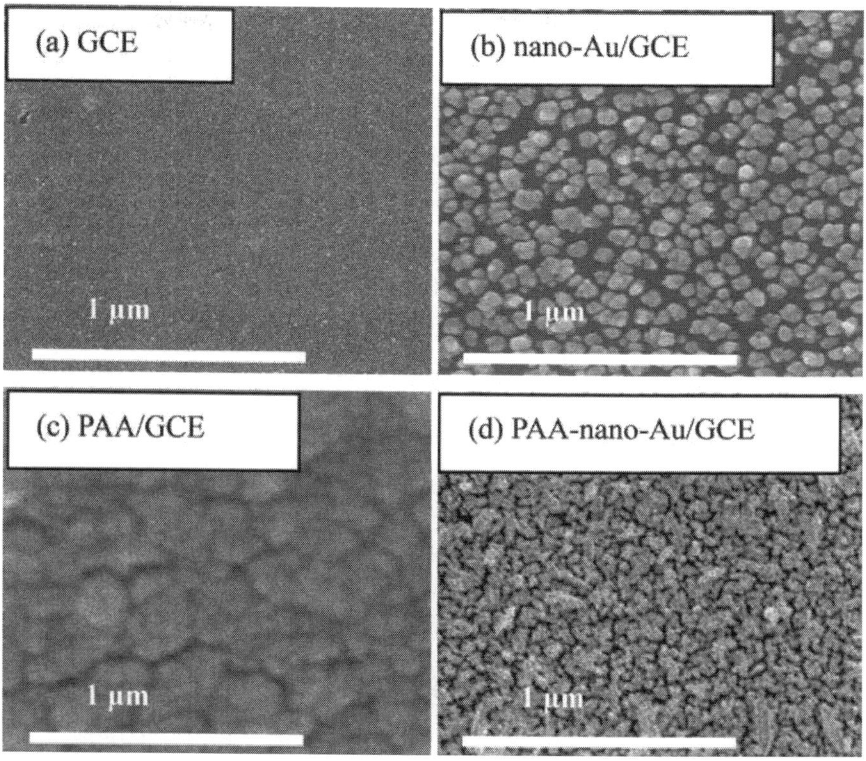

Figure 2. SEM image of (a) Bare GCE; (b) Nano-gold/GCE; (c) Poly Aspartic Acid/GCE; (d) Poly(aspartic acid)-nanogold/ GCE.

The electrochemical impedance spectroscopy (EIS) is an effective method to probe the surface features of the modified electrodes. The Nyquist plot of EIS includes a semicircle portion and a linear portion, with the former at higher frequencies corresponding to the electron-transfer limited process and the latter at lower frequencies corresponding to the diffusion process. The electron-transfer resistance (R_{ct}) at the electrode surface is equal to the semicircle diameter, which can be used to describe the interface properties of the electrode [33].

Figure 3 displays the EIS of bare GCE (a), PAA/GCE (b), nano-Au/GCE (c) and PAA-nano-Au/GCE (d), respectively, in 0.10 M KNO_3 solution containing 5.0 mM $K_3Fe(CN)_6/K_4Fe(CN)_6$ with the frequencies range from 0.1 to 1,00,000 Hz. As shown, the semicircle diameters decrease gradually from Figure 3(a) to (d), which indicates the electron transfer resistance at the electrode/ electrolyte interface decrease gradually. On the other hand, gold nanoparticles can act as electron transfer medium and enhance electron transfer and the result also confirms that the poly aspartic acid film incorporated gold nanoparticles had been successfully fabricated onto the glassy carbon electrode.

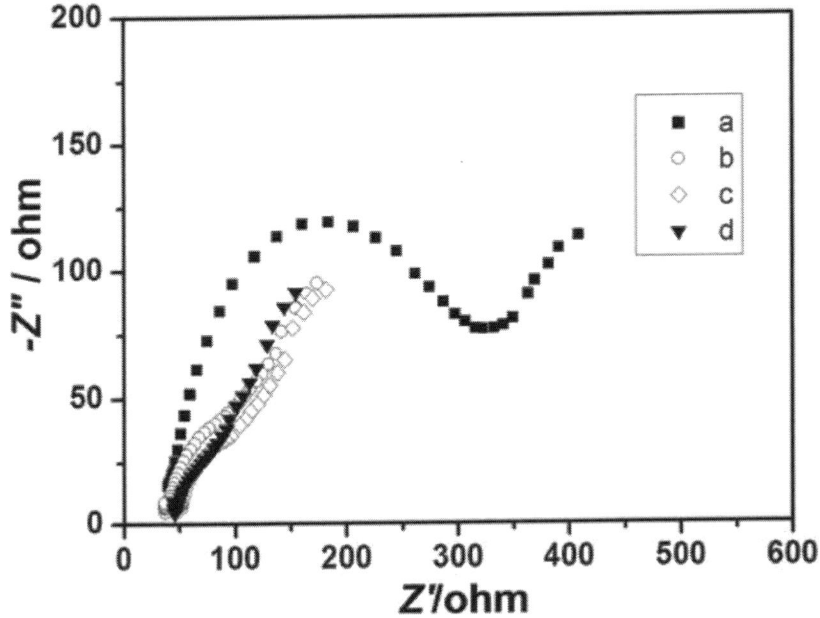

Figure 3. Electrochemical impedance spectroscopy (EIS) of: (a) Bare GCE; (b) Poly aspartic acid/GCE; (c) Nano-gold/ GCE; (d) Poly(aspartic acid)-nanogold/GCE, respectively in 0.10 M KNO_3 solution containing 5.0 mM $K_3Fe(CN)_6$/ $K_4Fe(CN)_6$.

3.2. Electrochemical Behavior of DA, AA and UA

The cyclic voltammograms of DA, AA and UA at bare GCE (a), nano-Au/GCE (b), PAA/GCE (c) and PAAnano-Au/GCE (d) in a phosphate buffer solution (pH 6.5) containing a ternary mixture of these compounds are shown in Figures 4(A)-(D) are CV recorded on bare GCE (dash line) and PAA-nano-Au/GCE (solid line) for DA, AA and UA respectively. As can be seen, the oxidation peaks appear at about 0.2 V, 0.19 V and 0.28 V at bare GCE for DA (B), AA (C) and UA (D), respectively. Bare GCE cannot clearly separate the oxidation peaks of AA and DA in the presence of UA, which showing two indistinguish and broad anodic peaks (Figure 4(A(a))). The phenomenon at nano-Au/GCE is similar to bare GCE (Figure 4(A(b))). Thus, it is unable to determine the three species simultaneously at the bare GCE or separate nano-Au modified electrode.

However, PAA-nano-Au/GCE has significant electrocatalytic activity towards the oxidation reactions of DA, AA and UA, and their oxidation peaks move to about 0.2 V (B), 0.0V (C) and 0.35 V (D), respectively.

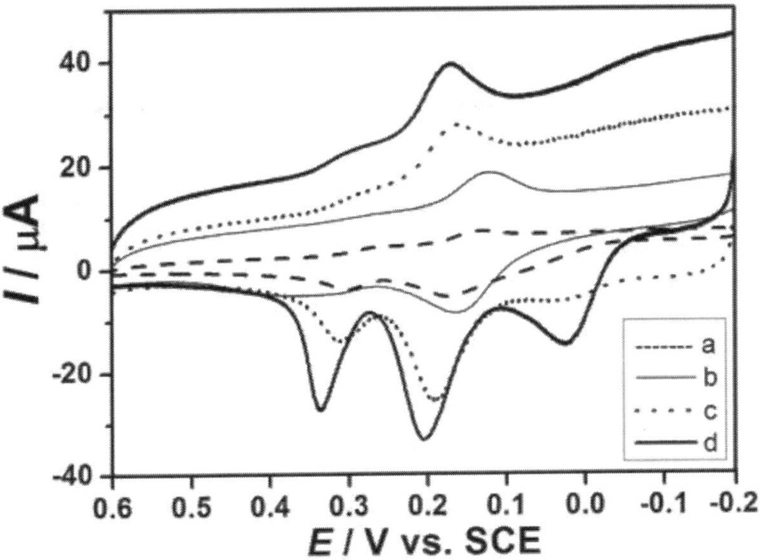
I / μA
E / V vs. SCE
a
b
c
d
40
20
0
-20
-40
0.6
0.5
0.4
0.3
0.2
0.1
0.0
-0.1
-0.2

(A)

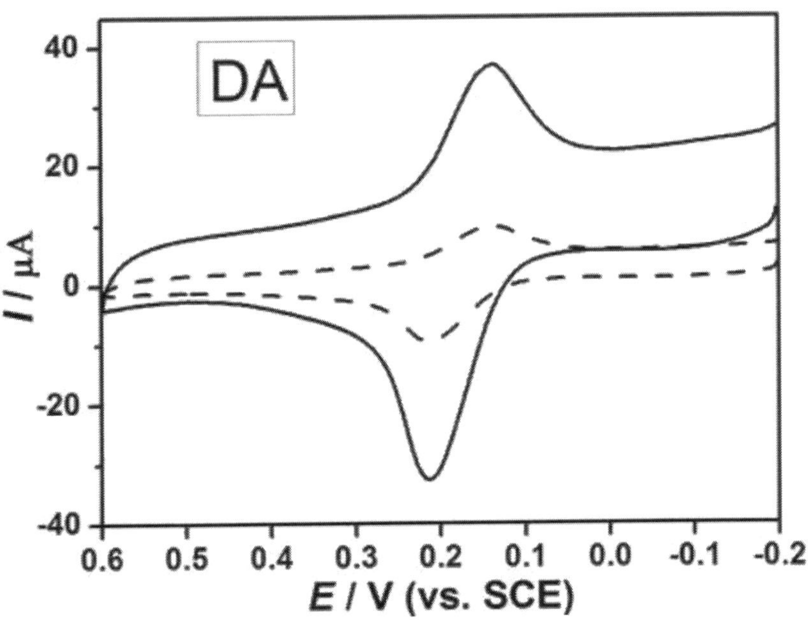
DA
I / μA
E / V (vs. SCE)
40
20
0
-20
-40
0.6
0.5
0.4
0.3
0.2
0.1
0.0
-0.1
-0.2

(B)

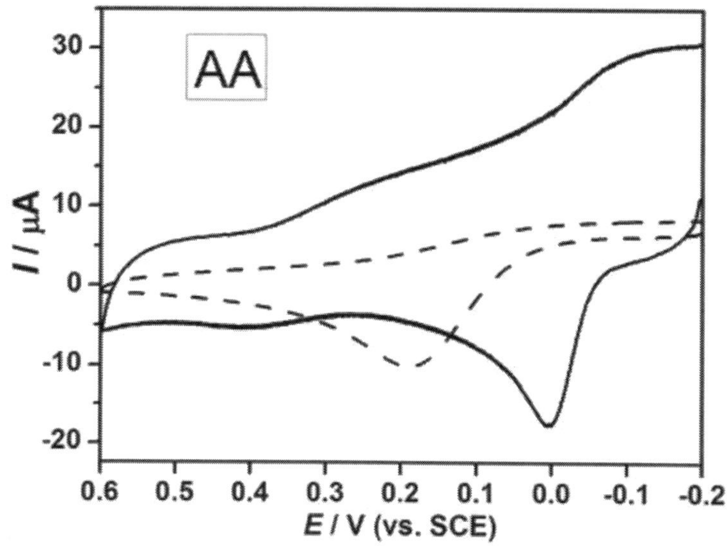

(C)

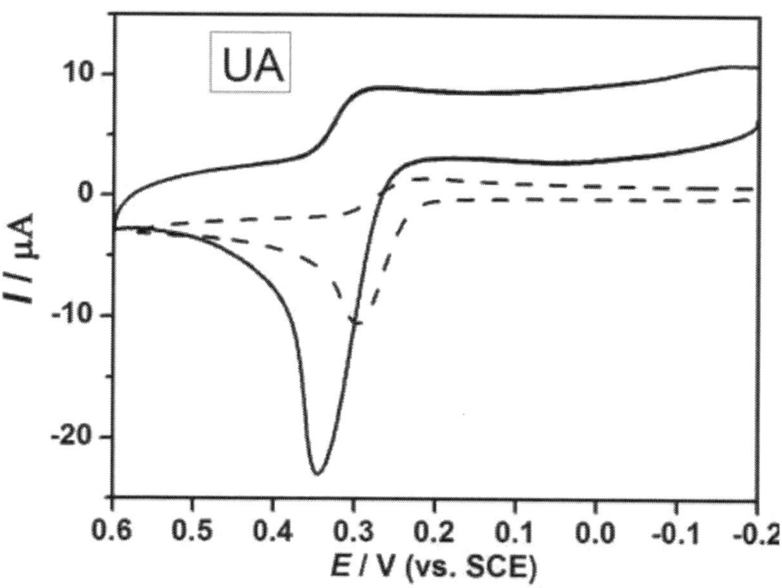

(D)

Figure 4. (A) Cyclic voltammograms of 100.0 μM UA, 50.0 μM DA and 250.0 μM AA at: (a) Bare GCE; (b) Nano-gold/ GCE; (c) Poly aspartic acid/GCE; (d) Poly(aspartic acid)- nanogold/GCE in buffer solution (Ph = 6.5) with scan rate of 100 $mV \cdot s^{-1}$. (B), (C) and (D) were CV recordeds on bare GCE (dash line) and PAA-nano-Au/GCE (solid line) for DA, AA and UA respectively.

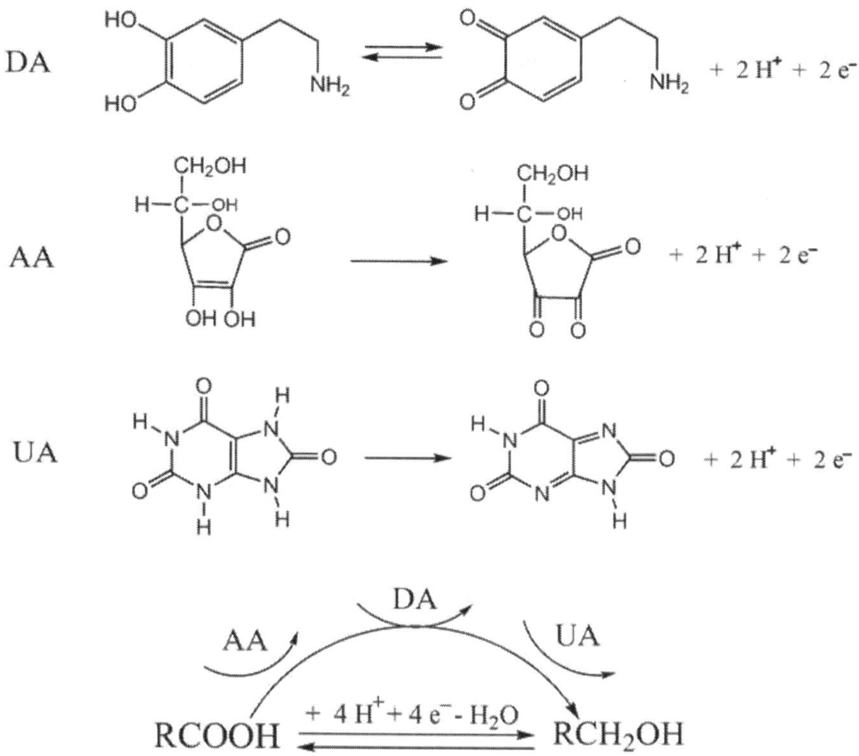

Scheme 2. Electrochemical oxidation reactions of DA, AA, and UA at poly(aspartic acid)-nanogold/GCE.

Also it can be seen that three well-defined oxidation peaks are observed (Figure 4(A(d))), the oxidation peak potential difference (ΔE_p) between AA and DA is 200 mV, DA and UA is 150 mV, which are large enough separations to allow for the simultaneous determination of DA, AA, and UA in a mixture. Compared with the PAA/GCE (Figure 4(A(c))), the peak currents (I_P) increased significantly due to the catalytic effect of gold nanoparticles.

In accordance with the literature [29,34], the mechanism for redox process of aspartic acid is a four-electron process. Electrochemical oxidation reactions of DA, AA, and UA at PAA-nano-Au/GCE are depicted in Scheme 2. The obviously enhanced current response and the decrease in the anodic overpotential of about 0.2 V for AA showed the strong catalytic function of PAA-nano-Au/ GCE to the oxidation of AA. The shift in the overpotentials was due to a kinetics effect, thus a substantial increase in the rate of electron transfer from AA was observed, which was attributed to the improvements in the reversibility of the electron transfer processes [35,36]. Because of the effect of both poly aspartic acid film and gold nanoparticles, there were some positive movement in the peak potentials for UA and no obvious change for DA, meanwhile the increased anodic peak currents is striking.

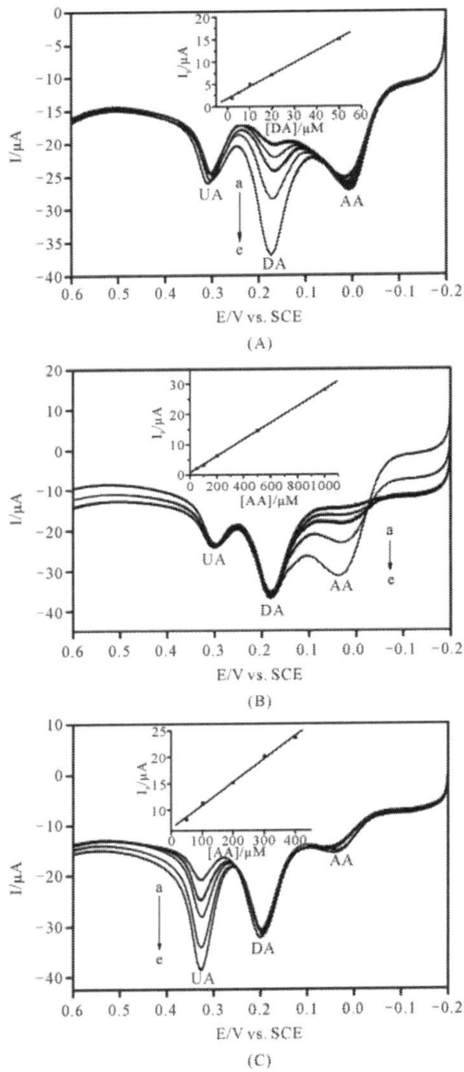

Figure 5. Voltammograms of (A): LSV of (a) 2.0; (b) 5.0; (c) 10; (d) 20 and (e) 50 μM DA in the presence of 50.0 μM UA and 100.0 μM AA; (insert graph is plot proportionate with these concentrations); (B): (a) 50; (b) 100; (c) 200; (d) 500 and (e) 1000 μM AA in the presence of 50.0 μM DA and 50.0 μM UA; (insert graph is plot proportionate with these concentrations); (C): (a) 50; (b) 100; (c) 200; (d) 300 and (e) 400 μM UA in the presence of 50.0 μM DA and 100.0 μM AA; (insert graph is plot proportionate with these concentrations) at pH 6.5 using the poly(aspartic acid)-nanogold/ GCE with scan rate of 100 mV·s^{-1}.

3.3. Simultaneous Determination of DA, AA, and UA

Figure 5 represents the linear sweep voltammograms (LSV) recordings of the solutions containing DA, AA and UA. It can be found that the peak height of each component is proportional to its concentration in the presence of the others. The linear relationships between the peak height and the concentration are obtained in the range of 5×10^{-7} - 1×10^{-4} M for DA, 5×10^{-6} - 2×10^{-3} M for AA and 5×10^{-6} - 1×10^{-3} M for UA, respectively. Detection limits of 6.5×10^{-8} M DA, 5.6×10^{-7} M AA and 3×10^{-7} M UA were obtained (based on S/N = 3). These results also confirm that the oxidation processes of DA, AA and UA at the PAA-nano-Au/GCE are independent from each other, which help the simultaneous determination of these three compounds without any interference.

To ascertain the reproducibility of the PAA-nano-Au/ GCE, three same GCE were modified to obtain the PAAnano-Au/GCE, and their responses towards DA, AA and UA were tested. The results indicate that resolution between the voltammetric signals of DA, AA and UA and the sensitivities are the same for all three PAA-nano-Au/ GCE. Besides, the long-term storage stability of PAAnano-Au/GCE was studied by keeping the electrode in 0.1 M phosphate buffer at room temperature. The response of the electrode in the solution containing 5×10^{-5} M DA, 1×10^{-4} M AA and 1×10^{-4} M UA, maintained 95%, 89%, 92% of the original peak current value after 7 days for DA, AA and UA, respectively. Such a good stability and reproducibility is acceptable for most practical applications.

3.4. Interference Study

The possible interferences of some inorganic salts and other organic compounds were also tested with 50.0 μM DA, 100.0 μM AA and 100.0 μM UA. The results indicate that five hundred-fold excess (vs DA) of K^+, Na^+, Mg^{2+}, NO_3^-, SO_4^{2-}, Cl^- and a hundred-fold excess (vs DA) organic compounds such as L-Glutamic acid, tartaric acid, citric acid, glucose, ethanol, and acetone have no influence on the signals of the three compounds with deviations less than 5%.

3.5. Application to Real Samples

We examined the applicability of the PAA-nano-Au/ GCE by measuring UA, DA and AA in calf serum and fetal calf serum samples. The calf serum and fetal calf serum samples were diluted five times with phosphate buffer solution before the measurements to prevent the matrix effect of real samples. The results are summarized in Table 2. The recovery of the samples ranged between 92.5% and 103.8%, indicating the detection procedures are free from interferences of the serum samples matrix.

Table 2. Determination of DA, AA and UA in real samples.

	Sample 1			Sample 2		
	UA	DA	AA	UA	DA	AA
Sample	26.75	-	-	19.99	-	-
Added (10^{-6} M)	100.0	50.0	500.0	100.0	50.0	500.0
Found[a] (10^{-6} M)	119.2 (±2.1)	51.9 (±0.7)	491.8 (±4.3)	14.8 (±1.7)	51.7 (±0.5)	477.5 (±4.6)
Recovery (%)	92.5	103.8	98.4	93.1	103.4	95.5

[a]Average values of three measurements.

4. CONCLUSIONS

The modification of the GCE surface by poly aspartic acid incorporated gold nanoparticles not only improves the electrochemical catalytic activities towards the oxidation of DA, AA and UA, but also resolves the overlapped oxidation peaks of DA, AA and UA into three well-defined peaks at potentials 200, 0, and 350 mV in CV, respectively. It also demonstrates that PAA-nano-Au/GCE facilitates the simultaneous determination of DA, AA and UA with good stability, sensitivity and selectivity compared with PAA/GCE or nano-Au/GCE. Due to the electrode's simple fabrication procedure, wide linear range, low detection limit for detection of DA, AA and UA as well as high stability and good reproducibility, this PAAnano-Au/GCE could be applied to the simultaneous determination of DA, AA and UA concentrations in real samples with satisfactory results.

5. ACKNOWLEDGEMENTS

This project was supported by the National Natural Science Foundation of China (Grant No. 20975061), the National Basic Research Program of China (Grant No. 2007CB936602), the Natural Science Foundation of Shandong Province in China (Grant No. Y2008B20).

REFERENCES

1. R. P. Da Silva, A. W. O. Lima and S. H. P. Serrano, "Simultaneous Voltammetric Detection of Ascorbic Acid, Dopamine and Uric Acid Using a Pyrolytic Graphite Electrode Modified Into Dopamine Solution," Analytica Chimca Acta, Vol. 612, No. 1, 2008, pp. 89-98.

2. X. Zhou, N. Zheng, S. R. Hou, X. J. Li and Z. B. Yuan, "Selective Determination Of Dopamine in the P-Resence of Ascorbic Acid at a Multi-Wall Carbon Nano-TubePoly(3,5-Dihydroxy Benzoic Acid) Film

Modified Electrode," Journal of Electroanalytical Chemistry, Vol. 642, No. 1, 2010, pp. 30-34.

3. T. J. Yin, W. Z. Wei and J. X. Zeng, "Selective Detection of Dopamine in the Presence of Ascorbic Acid by Use of Glassy-Carbon Electrodes Modified with Both Polyaniline Film and Multi-Walled Carbon Nanotubes with Incorporated β-Cyclodextrin," Analytical and Bioanalytical Chemistry, Vol. 386, No. 7-8, 2006, pp. 2087-2094.

4. J. B. Raoof, A. Kiani, R. Ojani, R. Valiollahi and S. R. Nadimi, "Simultaneous Voltammetric Determination of Ascorbic Acid and Dopamine at the Surface of Electrodes Modified with Self-Assembled Gold Nanoparticle Films," Journal of Solid State Electrochemistry, Vol. 14, No. 7, 2010, pp. 1171-1176.

5. O. Arrigoni and M. C. De Tullio, "Ascorbic Acid: More than Just an Antioxidant," Biochimica et Biophysica Acta, Vol. 1569, No. 1-3, 2002, pp. 1-9.

6. V. V. S. Eswara Dutt and H. A. Mottola, "Determination of Uric Acid at the Microgram Level by a Kinetic Procedure Based on a Pseudo-Induction Period," Analytical Chemistry, Vol. 46, No. 12, 1974, pp. 1777-1781.

7. W. N. Hu, D. M. Sun and W. Ma, "Silver Doped Poly(L-Valine) Modified Glassy Carbon Electrode for the Simultaneous Determination of Uric Acid, Ascorbic Acid and Dopamine," Electroanalysis, Vol. 22, No. 5, 2010, pp. 584-589.

8. C. F. Tang, S. A Kumar and S. M. Chen, "Zinc Oxide/Redox Mediator Composite Films-Based Sensor for Electrochemical Detection of Important Biomolecules," Analytical Biochemistry, Vol. 380, No. 2, 2008, pp. 174- 183.

9. J. S. Huang, Y. Liu, H. Q. Hou and T. Y. You, "Simultaneous Electrochemical Determination of Dopamine, Uric Acid and Ascorbic Acid Using Palladium NanoparticleLoaded Carbon Nanofibers Modified Electrode," Biosensors and Bioelectronics, Vol. 24, No. 4, 2008, pp. 632- 637.

10. A. Salimi, H. MamKhezri and R. Hallaj, "Simultaneous Determination of Ascorbic Acid, Uric Acid and Neurotransmitters with a Carbon Ceramic Electrode Prepared by Sol-Gel Technique," Talanta, Vol. 70, No. 4, 2006, pp. 823-832.

11. S. Thiagarajan and S. M. Chen, "Preparation and Characterization of Ptau Hybrid Film Modified Electrodes and Their Use in Simultaneous Determination of Dopamine, Ascorbic Acid and Uric Acid," Talanta, Vol. 74, No. 2, 2007, pp. 212-222.

12. A. Safavi, N. Maleki, O. Moradlou and F. Tajabadi, "Simultaneous Determination of Dopamine, Ascorbic Acid, and Uric Acid Using Carbon Ionic Liquid Electrode," Analytical Biochemistry, Vol. 359, No. 2, 2006, pp. 224- 229. doi:10.1016/j.ab.2006.09.008

13. P. Shakkthivel and S. M. Chen, "Simultaneous Determination of Ascorbic Acid and Dopamine in the Presence of Uric Acid on Ruthenium Oxide Modified Electrode," Biosensors and Bioelectronics, Vol. 22, No. 8, 2006, pp. 1680-1687.

14. Y. X. Li and X. Q. Lin, "Simultaneous Electroanalysis of Dopamine, Ascorbic Acid and Uric Acid By Poly(Vinyl Alcohol) Covalently Modified Glassy Carbon Electrode," Sensors and Actuators B: Chemical, Vol. 115, No. 1, 2006, pp. 134-139.

15. L. Q. Lin, J. H. Chen, H. Yao, Y. Z. Chen, Y. J. Zheng and X. H. Lin, "Simultaneous Determination of Dopamine, Ascorbic Acid and Uric Acid at Poly(Evans Blue) Modified Glassy Carbon Electrode," Bioelectrochemistry, Vol. 73, No. 1, 2008, pp. 11-17.

16. H. R. Zare, N. Rajabzadeh, N. Nasirizadeh and M. M. Ardakani, "Voltammetric Studies of an Oracet Blue Modified Glassy Carbon Electrode and Its Application for the Simultaneous Determination of Dopamine, Ascorbic Acid and Uric Acid," Journal of Electroanalytical Chemistry, Vol. 589, No. 1, 2006, pp. 60-69.

17. X. H. Lin, Q. Zhuang, J. H. Chen, S. B. Zhang and Y. J. Zheng, "Electrocatalytic Property of Poly-Chro-Motrope 2B Modified Glassy Carbon Electrode on Dopamine and Its Application," Sensors and Actuators B: Chemical, Vol. 125, No. 1, 2007, pp. 240-245.

18. J. H. Chen, J. Zhang, X. H. Lin, H. Y. Wan and S. B. Zhang, "Electrocatalytic Oxidation and Determination of Dopamine in the Presence of Ascorbic Acid and Uric Acid at a Poly (4-(2-Pyridylazo)-Resorcinol) Modified Glassy Carbon Electrode," Electroanalysis, Vol. 19, No. 5, 2007, pp. 612-619.

19. H. Yao, Y. Y. Sun, X. H. Lin, Y. H. Tang and L. Y. Huang, "Electrochemical Characterization of Poly(Eriochrome Black T) Modified Glassy Carbon Electrode and Its Application to Simultaneous Determination of Dopamine, Ascorbic Acid and Uric Acid," Electrochimica Acta, Vol. 52, No. 20, 2007, pp. 6165-6171.

20. A. A. Ensafi, M. Taei, T. Khayamian and A. Arabz-Adeh, "Highly Selective Determination of Ascorbic Acid, Dopamine, and Uric Acid by Differential Pulse Volta-Mmetry Using Poly(Sulfonazo III) Modified Glassy Carbon Electrode," Sensors and Actuators B: Chemical, Vol. 147, No. 1, 2010, pp. 213-221.

21. P. R. Roy, T. Okajima and T. Ohsaka, "Simultaneous Electroanalysis of Dopamine and Ascorbic Acid Using Poly(N, N-Dimethylaniline)-Modified Electrodes," Bioelectrochemistry, Vol. 59, No. 1-2, 2003, pp. 11-19.

22. V. S. Vasantha and S. M. Chen, "Electrocatalysis and Simultaneous Detection of Dopamine and Ascorbic Acid Using Poly(3,4-Ethylenedioxy) Thiophene Film Modified Electrodes," Journal of Electroanalytical Chemistry, Vol. 592, No. 1, 2006, pp. 77-87.

23. U. Yogeswaran and S. M. Chen, "Multi-Walled Carbon Nanotubes with Poly(Methylene Blue) Composite Film for the Enhancement and Separation of Electroanalytical Responses of Catecholamine and Ascorbic Acid," Sensors and Actuators B: Chemical, Vol. 130, No. 2, 2008, pp. 739-749.

24. A. Balamurugan and S. M. Chen, "Poly(3,4-Ethyl-enedioxythiophene-co-(5-amino-2-naphthalenesulfonic acid)) (PEDOT-PANS) Film Modified Glassy Carbon Electrode for Selective Detection of Dopamine in the Presence of Ascorbic Acid and Uric Acid," Analytica Chimica Acta, Vol. 596, No. 1, 2007, pp. 92-98.

25. P. Kalimuthu and S. A. John, "Electropolymerized Film of Functionalized Thiadiazole on Glassy Carbon Electrode for the Simultaneous Determination of Ascorbic Acid, Dopamine and Uric Acid," Bioelectrochemistry, Vol. 77, No. 1, 2009, pp. 13-18.

26. L. Han and X. L. Zhang, "Simultaneous Voltammetry Determination of Dihydroxybenzene Isomers by Nanogold Modified Electrode," Electroanalysis, Vol. 21, No. 2, 2009, pp. 124-129.

27. B. Serra, S. Jiménez, M. L. Mena, A. J. Reviejo and J. M. Pingarrón, "Composite Electrochemical Biosensors: A Comparison of Three Different Electrode Matrices for the Construction of Amperometric Tyrosinase Biosensors," Biosensors and Bioelectronics, Vol. 17, No. 3, 2002, pp. 217-226.

28. P. Santhosh, A. Gopalan and K. P. Lee, "Gold Nanoparticles Dispersed Polyaniline Grafted Multiwall Carbon Nanotubes as Newer Electrocatalysts; Preparation and Performances for Methanol Oxidation," Journal of Catalysis, Vol. 238, No. 1, 2006, pp. 177-185.

29. L. Zhang and X. Q. Lin, "Electrochemical Behavior of a Covalently Modified Glassy Carbon Electrode with Aspartic Acid and Its Use for Voltammetric Differentiation of Dopamine and Ascorbic Acid," Analytical and Bioanalytical Chemistry, Vol. 382, No. 7, 2005, pp. 1669-1677.

30. U. Yogeswaran and S. M. Chen, "Separation and Concentration Effect of f-MWCNTs on Electrocatalytic Responses of Ascorbic Acid, Dopamine and Uric Acid at f-MWCNTs Incorporated with Poly(Neutral Red) Composite Films," Electrochimica Acta, Vol. 52, No. 19, 2007, pp. 5985-5996.

31. A. J. Downard and A. B. Mohamed, "Suppression of Protein Adsorption at Glassy Carbon Electrodes Covalently Modified with Tetraethylene Glycol Diamine," Electroanalysis, Vol. 11, No. 6, 1999, pp. 418-423.

32. R. S. Deinhammer, H. Mankit, J. W. Anderegg and M. D. Porter, "Electrochemical Oxidation of Amine-Containing Compounds: A Route to the Surface Modification of Glassy Carbon Electrodes," Langmuir, Vol. 10, No. 4, 1994, pp. 1306-1313.

33. G. Zhao, J. J. Xu and H. Y. Chen, "Fabrication, Characterization of Fe_3O_4 Multilayer Film and Its Application in Promoting Direct Electron Transfer of Hemoglobin," Electrochemistry Communication, Vol. 8, No. 1, 2006, pp. 148-154.

34. H. X. Luo, Z. J. Shi, N. Q. Li, Z. N. Gu and Q. K. Zhuang, "Investigation of the Electrochemical and Electrocatalytic Behavior of Single-Wall Carbon Nanotube Film on a Glassy Carbon Electrode," Analytical Chemistry, Vol. 73, No. 5, 2001, pp. 915-920.

35. F. Malem and D. Mandler, "Self-Assembled Monolayers in Electroanalytical Chemistry: Application of Omegamercapto Carboxylic Acid Monolayers for the Electrochemical Detection of Dopamine in the Presence of a High Concentration of Ascorbic Acid," Analytical Chemistry, Vol. 65, No. 1, 1993, pp. 37-41.

36. L. Zhang and X. Q. Lin, "Covalent Modification of Glassy Carbon Electrode with Glutamic Acid for Simultaneous Determination of Uric Acid and Ascorbic Acid," Analyst, Vol. 126, No. 3, 2001, pp. 367-370.

Index